Artificial Intelligence and Data Mining Approaches in Security Frameworks

Scrivener Publishing
100 Cummings Center, Suite 541J
Beverly, MA 01915-6106

Advances in Cyber Security

Series Editor: Rashmi Agrawal and D. Ganesh Gopal

Scope: The purpose of this book series is to present books that are specifically designed to address the critical security challenges in today's computing world including cloud and mobile environments and to discuss mechanisms for defending against those attacks by using classical and modern approaches of cryptography, blockchain and other defense mechanisms. The book series presents some of the state-of-the-art research work in the field of blockchain, cryptography and security in computing and communications. It is a valuable source of knowledge for researchers, engineers, practitioners, graduates, and doctoral students who are working in the field of blockchain, cryptography, network security, and security and privacy issues in the Internet of Things (IoT). It will also be useful for faculty members of graduate schools and universities. The book series provides a comprehensive look at the various facets of cloud security: infrastructure, network, services, compliance and users. It will provide real-world case studies to articulate the real and perceived risks and challenges in deploying and managing services in a cloud infrastructure from a security perspective. The book series will serve as a platform for books dealing with security concerns of decentralized applications (DApps) and smart contracts that operate on an open blockchain. The book series will be a comprehensive and up-to-date reference on information security and assurance. Bringing together the knowledge, skills, techniques, and tools required of IT security professionals, it facilitates the up-to-date understanding required to stay one step ahead of evolving threats, standards, and regulations.

Publishers at Scrivener
Martin Scrivener (martin@scrivenerpublishing.com)
Phillip Carmical (pcarmical@scrivenerpublishing.com)

Artificial Intelligence and Data Mining Approaches in Security Frameworks

Edited by

Neeraj Bhargava
Ritu Bhargava
Pramod Singh Rathore
Rashmi Agrawal

Scrivener
Publishing

WILEY

This edition first published 2021 by John Wiley & Sons, Inc., 111 River Street, Hoboken, NJ 07030, USA and Scrivener Publishing LLC, 100 Cummings Center, Suite 541J, Beverly, MA 01915, USA
© 2021 Scrivener Publishing LLC
For more information about Scrivener publications please visit www.scrivenerpublishing.com.

Wiley Global Headquarters
111 River Street, Hoboken, NJ 07030, USA

For details of our global editorial offices, customer services, and more information about Wiley products visit us at www.wiley.com.

Limit of Liability/Disclaimer of Warranty
While the publisher and authors have used their best efforts in preparing this work, they make no representations or warranties with respect to the accuracy or completeness of the contents of this work and specifically disclaim all warranties, including without limitation any implied warranties of merchantability or fitness for a particular purpose. No warranty may be created or extended by sales representatives, written sales materials, or promotional statements for this work. The fact that an organization, website, or product is referred to in this work as a citation and/or potential source of further information does not mean that the publisher and authors endorse the information or services the organization, website, or product may provide or recommendations it may make. This work is sold with the understanding that the publisher is not engaged in rendering professional services. The advice and strategies contained herein may not be suitable for your situation. You should consult with a specialist where appropriate. Neither the publisher nor authors shall be liable for any loss of profit or any other commercial damages, including but not limited to special, incidental, consequential, or other damages. Further, readers should be aware that websites listed in this work may have changed or disappeared between when this work was written and when it is read.

Library of Congress Cataloging-in-Publication Data

ISBN 978-1-119-76040-5

Cover image: (Antenna Tower): Carmen Hauser | Dreamstime.com
Cover design by Kris Hackerott

Contents

Preface

Artificial Intelligence (AI) and data mining not only provide a better understanding of how real-world systems function, but they also enable us to predict system behavior before a system is actually built. They can also accurately analyze systems under varying operating conditions. This book provides comprehensive, state-of-the-art coverage of all the important aspects of modeling and simulating both physical and conceptual systems. Various real-life examples show how simulation plays a key role in understanding real-world systems. We also explained how to effectively use AI and Data Mining techniques to successfully apply the modeling and simulation techniques presented.

After introducing the underlying philosophy of systems, the book offers step-by-step procedures for modeling with practical examples and coding different types of systems using modeling techniques, such as the Pattern Recognition, Automatic Threat detection, Automatic problem solving, etc.

Preparing both undergraduate and graduate students for advanced modeling and simulation courses, this text helps them carry out effective simulation studies. In addition, graduate students should be able to comprehend and conduct AI and Data Mining research after completing this book.

This book is organized into fifteen chapters. In Chapter 1, this Chapter discusses about the cyber security needs that can be addressed by AI techniques. It talks about the traditional approach and how AI can be used to modify the multilayered security mechanism used in companies today. Here we propose a system that adds additional layer of security in order to detect any unwanted intrusion. The ever-expanding danger of digital assaults, cybercrimes, and malware attacks has grown exponentially with evolution of artificial intelligence. Conventional ways of cyber-attacks have now taken a turning point, consequently, the attackers resort to more intelligent ways.

In Chapter 2, we have tried to show the power of intrusion detection is the most important application of data mining by applying different data mining techniques to detect it effectively and report the same in actual time so that essential and required arrangements can be made to stop the efforts made by the trespassery.

In Chapter 3, we have explained about how Artificial Intelligence (AI) is a popular expression in the digital world. It is as yet a creating science in various features as indicated by the difficulties tossed by 21st century. Usage of artificial intelligence has gotten undefined from human life. Nowadays one can't imagine a world without AI as it has a ton of gigantic impact on human life. The essential objective of artificial intelligence is to develop the advancement based activities which addresses the human data in order to handle issues. Basically artificial intelligence is examination of how an individual think, work, learn and pick in any circumstance of life, whether or not it may be related to basic reasoning or learning new things or thinking equitably or to appear at an answer, etc.

In Chapter 4, we have explained further proposed a botnet identification version using optics algorithm that hopes to effectively discover botnets and perceive the type botnet detected by way of addition of latest feature; incorporation of changed traces to pinpoint supply IP of bot master, identification of existence of the kind of services the botnets have get right of entry to to are areas the proposed solution will cater for.

In Chapter 5, we have explained about models basically 'learns' from experience with respect to some task and are capable of finding 'commonality' in many different observations. This study discusses various methods of spam filtering using existing Artificial Intelligence techniques and compares their strengths and limitations.

In Chapter 6, we have explained about how as artificial intelligence people in general to improve, there are risks associated with their utilization, set up in functioning frameworks, tools, calculations, framework the executives, morals and duty, and privacy. The study focuses around the risks and threats of computerized reasoning and how AI can help comprehend network safety or areas of cyber security issues.

In Chapter 7, we have explained about problem to make privacy in multi-tenant in the single framework. For that problem we use the artificial intelligence concept to improve the security and privacy concept in multi-tenant based system. Using Artificial intelligence the privacy and security concept make strong because in artificial intelligence work as intelligent human or animal mind it make maximum changes to fulfill the requirement of the concept to achieve the goal. In this chapter describes the issues of privacy and security problems in multi tenancy.

In Chapter 8, we have provided detailed explanations of a novel approach for biometric recognition has been introduced in which the application of ILBP (Improved Local Binary Pattern) for facial feature detection is discussed which generates the improved features for the facial pattern. It allows only authenticated user to access a system which is better than previous algorithms. Previous research for face detection shows many demerits in terms of false acceptance and rejection rates. In this paper, the extraction of facial features is done from static and dynamic frames using the Haar cascade algorithm.

In Chapter 9, we have explained about a the developed system consists of a climbing robot, camera for image capturing, IoT modules for transmitting images to cloud, image processing platform, and artificial neural network module intended for decision making. Climbing robot holds the cable with the grooved wheels along with the auto trigger camera and the IoT module. For inspection, the robot ascends along the cables continuously and acquires images of various segments of the cable. Then the captured images have been send to the cloud storage through IoT system. The stored images have been retrieved and their sizes have been reduced through the image processing techniques.

In Chapter 10, we have a digital security threats results from the character of those omnipresent and at times over the top interchanges interconnections. Digital security isn't one aspect, yet rather it's a gaggle of profoundly various issues mentions various arrangements of threats. An Advance Cyber Security System utilizing emblematic rationale might be a framework that comprises of a standard safe and an instrument for getting to and running the standards. The vault is ordinarily built with a lot of related standard sets. Fuzzy improvement manages finding the estimations of information boundaries of a luxurious recreated framework which winds up in wanted yield.

In Chapter 11, the goal of current chapter is to analyze cyber threats and to demonstrate how artificial intelligence and data mining approaches can be effective to fix cyber-attack issues. The field of artificial intelligence has been increasingly playing a vital role in analyzing cyber threat and improving cyber security as well as safety. Mainly three aspects are discussed in this chapter. First the process of cyber-attack detection which will help to analyses and classify cyber incident, Second task is forecasting upcoming cyber-attack and to control the cyber terrorism. Finally the chapter focus on theoretical background and practical usability of artificial intelligence with data mining approaches for addressing above detection and prediction.

In Chapter 12, this chapter explores the modern intrusion detection with a distinctive determination perspective of data mining. This discussion focuses on major facets of intrusion detection strategy that is misuse detection. Below content focuses on, to identify attacks, information or data which is present on the network using C4.5 algorithm, which is type of decision tree technique and also it helps to enhance the IDS system to recognize types of attacks in network. For this attack detection, KDD-99 dataset is used, contains several features and different class of general and attack type data.

In Chapter 13, in this current research, firefly algorithm has been used for optimizing maize crop yield by considering the various constraints and risks. This research investigates the development of new firefly algorithm module for predicting the optimal climatic conditions and predicts the crop cultivation output. As the pre-processing, the maize crop cultivation data for 96 months have been collected and provided as response to Minitab software to formulate the relational equation. The collected data have been stored in the cloud using IoT and the cloud has to be updated periodically for obtaining the accurate results from the algorithm.

In Chapter 14, gestures are of two types as: static and dynamic sequences, this is where vision based techniques plays a vital role. The survey on the research study on the vision-based gesture recognition approaches have been briefed in this paper. Challenges in all perspective of recognition of gestures using images are detailed. A systematic review has been conducted over 100 papers and narrowed down into 60 papers on summarized. The foremost motive of this paper is to provide a strong foundation on vision based recognition and apply this for solutions in medical and engineering fields. Outlines gaps & current trends to motivate researchers to improve their contribution.

In Chapter 15, we will cover a examine of diverse thoughts, attempts, efficiency and different studies trends in junk mail filtering. The history observe explains the packages of device gaining knowledge of strategies to clear out the antispam emails of main e mail service carriers like gmail, yahoo, outlook and so on. We can talk the e-mail unsolicited mail filtering techniques and sundry efforts made via various researchers in fighting the unsolicited mail emails via using device mastering strategies. Here, we talk and make comparisons within the strengths & weaknesses of already present machine learning algorithms & techniques and different open studies troubles in spam filtering. We might suggest deep gaining knowledge & deep adversarial getting to know as these technologies are the destiny to be able to capable of efficaciously deal with spam emails threats.

Prof. Neeraj Bhargava
Professor & Head
Department of Computer Science
School of Engineering and System Science
MDS University, Ajmer, Rajasthan, India

Dr. Ritu Bhargava
Assistant Professor
Department of Computer Science
Sophia Girl's College Autonomous
Ajmer, Rajasthan, India

Pramod Singh Rathore
Assistant Professor
Aryabhatta College of Engineering and Research Center,
Ajmer, Rajasthan, India
Department of Computer Science & Engineering
Visiting Faculty, MDS University, Ajmer, Rajasthan, India

Prof. Rashmi Agrawal
Professor
Manavrachna International Institute of Research and Studies,
Faridabad, India

Role of AI in Cyber Security

Navani Siroya[1]* and Prof Manju Mandot[2]

[1]M.Tech Scholar, Computer Science, MDS University, Ajmer, India
[2]Director, Department of Computer Science and IT JRN Rajasthan Vidyapeeth
University, Udaipur, Rajasthan

Abstract

Borderless cyberspace as a part of "global commons" does not exist. Information breaches, ID theft, cracking the captcha, and other such stories proliferate, more so in times of the pandemic era, affecting people at a global level.

Today where AI advancements, for example, deep learning, can be incorporated into cyber security to develop shrewd models for executing malware classification, intrusion detection and threatens intelligence sensing. The flip side of the coin shows how AI models have to confront different digital dangers, which upsets their sampling, learning models, and decision-making. The ever-expanding danger of digital assaults, cybercrimes, and malware attacks has grown exponentially with the evolution of artificial intelligence. Conventional ways of cyber-attacks have now taken a turning point; consequently, the attackers resort to more intelligent ways.

This chapter discusses the cyber security needs that can be addressed by AI techniques. It talks about the traditional approach and how AI can be used to modify the multilayered security mechanism used in companies today. Here we propose a system that adds an additional layer of security in order to detect any unwanted intrusion. The chapter ends with deliberations on the future extent of artificial intelligence and cyber security.

Keywords: Artificial intelligence, cyber security, machine learning, Botnet

*Corresponding author: siroyanavani@gmail.com

Neeraj Bhargava, Ritu Bhargava, Pramod Singh Rathore, and Rashmi Agrawal (eds.) Artificial Intelligence and Data Mining Approaches in Security Frameworks, (1–10) © 2021 Scrivener Publishing LLC

1.1 Introduction

Artificial Intelligence (AI) can be characterized as artificial decision making similar to human decision making, based on certain unique algorithms and related mathematical estimations. Cyber Security relates to measures taken to protect against digital assaults in the virtual world.

Moreover, the job of AI is ever expanding in the modern world, where there is a looming threat to cyber security.

With the headway in innovation, cybercrimes are also increasing and getting unpredictable. Cyber criminals are launching sophisticated attacks that are putting current security frameworks in danger. Thus, the cyber security business is evolving to satisfy the expanding security needs of organizations. But, these defensive strategies of security professionals may not live up expectations and may fall short of its proposed agenda sooner or later [1].

1.2 Need for Artificial Intelligence

AI's vital job is to offload work from human cyber security engineers presently, to deal with the depth and detail that humans cannot tackle effectively. Advancement in machine learning technology implies that AI applications can also automatically adapt to changes in threats and spot issues as they emerge.

Cyber security needs that AI tools and platforms can help to meet:

> Data Extent
> People get confused immediately when confronted with huge amounts of log information and cautions delivered by the present frameworks. Simulated intelligence programming running on today's powerful processors can go through more data in minutes than humans could handle in months. Thus, it can also account for issues and inconsistencies while taking care of enormous volumes of security information.

> Threat needles
> Cyber threat hunting is a constant proactive search through networks and data sets to detect threats that elude existing computerized tools.
> Digital lawbreakers are now inside numerous frameworks, waiting to complete their attacks. They can frequently escape

people. But AI can quickly examine different circumstances to detect the threat needles compared to malicious activity.

Optimization of Response
Artificial intelligence can accelerate recognition of certifiable issues, quickly cross-referencing various alerts and sources of security information. The priorities of the incidents to be dealt with will still be the domain of human cyber security experts but they can be further helped by AI systems that will increase speed of recognition and reaction times.

AI arms race
Cyber criminals today are already equipped with advanced AI techniques. AI technology in general can be a boon or a bane. Programmers can easily utilize the most recent tools to launch more sophisticated attacks, each one being more dangerous. It has become an arms race where AI is the main exponent on both sides [2, 3].

1.3   Artificial Intelligence in Cyber Security

AI in cyber security supports companies or organizations, allowing them to safeguard their defense mechanisms; furthermore, it helps them to interpret cybercrimes effectively. Enterprises are using this ideal opportunity to achieve efficiency in automation by going digital as they take leverage of faster execution speeds. Achieving digital connectedness in their entire value chains helps them to meet the increasing competition in the market. On similar lines, cybercriminals find opportunity with increasing digitization. Cybercrime unions are actively focusing on digital ecosystems including cloud infrastructure, Internet of Things (IoT) devices and software as a service (SaaS) offerings. Therefore, Enterprises are confronted with the challenge of pushing for greater gains in business advantage while balancing the risk of cyber exposure [4, 5].

1.3.1   Multi-Layered Security System Design

Organizations are concentrating more on cyber security in the present scenario. This is because advanced cyber security attacks have forced them to spend a lot of money to prevent future data breaches. It begins with designing a multi-layered security framework that will secure the network infrastructure.

Figure 1.1 Network infrastructure [4].

Figure 1.1 denotes the network infrastructure which contain Firewall, anti-virus software and a disaster recovery plan. All these component make a network infrastructure more efficient. AI has affected security by helping experts to recognize abnormalities in the system by analyzing client activities and contemplating the examples. Security experts would now be able to contemplate and organize information utilizing AI and detect vulnerabilities to forestall harmful attacks [6].

1.3.2   Traditional Security Approach and AI

AI will help enhance the traditional security approach in the following ways:

- Advanced AI-powered security instruments will be utilized to screen and react to security events.
- Modern firewalls will have built-in machine learning technology to detect and remove an unusual pattern in the system traffic, if considered hostile.
- Analyzing vulnerabilities using the natural language processing feature in AI, security experts can also identify the root of a digital assault.
- Predictive analysis of detecting malicious threats and scanning of the data in advance is required.

Since our reliance on big data has increased, we have created a parallel need of keeping it safe. Thus, the need of the hour is to safeguard the integrity of networks, stored data and programs from unauthorized access and attacks [7–9].

1.4 Related Work

1.4.1 Literature Review

Today the Internet is used by millions of ordinary people, making them easy targets for cyber criminals. With "softwarisation and digitization" and rapid adoption of IoT, cyber security is now at the core of business strategy. Data is a broad categorization, ranging from credit-card information, bank, financial records and personal information. The contemporary solution to this far-reaching issue lies in basic awareness, building defensive cyber-capabilities or protection and care, by education.

Onashoga, S. Adebukola, Ajayi, O. Bamidele and A. Taofik (2013) in their paper discuss simulated multi-agent-based architecture for intrusion detection system to overcome the shortcoming of the current mobile agent–based intrusion detection system. The data are distributed on both the host and the network. Closed pattern mining (CPM) algorithm is introduced for profiling the users' activities in network database. This not only helps in reducing the time of sorting the data but also helps the analysts to know about the patterns of human behavior in real time.

Alex Roney Mathew *et al.* (2010) in their paper discuss the different types of cybercrime, namely: social-engineering phishing, email spoofing and pharming. They also discuss ways to protect people from such crimes with an emphasis on biometrics. Cybercrimes have lately become so prevalent in that only a very small amount of the population in the world has been untouched by them [1].

Selvakani, Maheshwari V. and Karavanisundari (2010) in their paper stress the fact that information technology can be used for destructive as well as constructive work, depending on whose hands it ends up in. The study deals with the importance of cyber laws to protect the interests of the cyber victims. The authors believe that a computer can be secured even by a person with simple knowledge but that the ascertainment and preservation of the evidence is a tough task. There is a need for techno-legal harmonized law; a good combination is required. AI should help in designing a strong law which can be used effectively to trace cybercrimes.

L.S. Wijesinghe, L.N.B. De Silva, G.T.A. Abhayaratne, P. Krithika, S.M.D.R. Priyashan, and Dhishan Dhammearatchi (2016) in their research paper mainly focus on how to combat cybercrimes, and also throw some light on how intelligent and effective the tool "agent" can be used in the detection and prevention of cyber-attacks. Cyber-attacks tend to have a huge impact on the IT industry when it comes to data theft, the data has become more vulnerable and prone to such threats and attacks due to the malicious activities going on for the agendas best known to the attackers.

Ramamoorthy R. (2010) in his paper discusses the various perspectives of cyber security. Because of ontinuously evolving new threats against enterprises, IT has made cyber security a "must look-into" issue. The systems administration team should devise ways to improve their cyber security with an automated, on-demand, real-time application security testing solution that makes comprehensive cyber security for applications simpler and more cost effective. Cyber security knows no borders. The author touches upon controlling server sprawl to increase operational efficiency and ease disaster recovery, virtualization clearly delivers end results.

Yasmin N., and Bajaj N. (2012) in their research paper present S-box Modification in DES. DES is Data Encryption Standards and S-box "substitution box" – a standard encryption device. Security is the main concern for organizations participating in information exchange. One of the essential aspects for secure communications is that of cryptography. As cybercrimes are causing serious financial losses, an existing system needs constant modifications to ensure that security levels are not compromised. It shows a higher degree of resistance against attack on relationship $L_{i+1} = R_i$. But a significant amount of mathematical knowledge and understanding the complete crypto system is required.

1.4.2   Corollary

Cyber-attacks are expanding quickly, notwithstanding increased security measures. The attacks may be a malware, phishing attack, password theft, Trojan attacks, and so on. In order to avoid these cybercrimes robust cyber-security measures are needed. Emerging technologies like cognitive science, cloud computing, robotics, internet banking, and e-commerce urgently need to put in adequate safeguard measures in the domain of cyber security. With increasing use of Big Data, governing the decision-making by use of machine learning models, cyber security needs to be at the helm [10, 11].

1.5   Proposed Work

The model proposes to add a layer of security to the multi-layered security approach. The proposed system architecture describe in Figure 1.2.

1. Suppose, while we are trying to log into our bank account using our credentials, a bot tries to crack the captcha.

2. Whenever it does so a machine learning model based on its ability to recognize patterns from the past would detect the presence of bots through active monitoring and predictive analysis.
3. If detected, it would terminate the current process and send out an alert.
4. If the bot is not present then it would continue the process and run the anti-virus software, in order to remove any other malicious files.
5. The Disaster recovery plan in the end would ensure that any important data is not lost and is backed up.

1.5.1 System Architecture

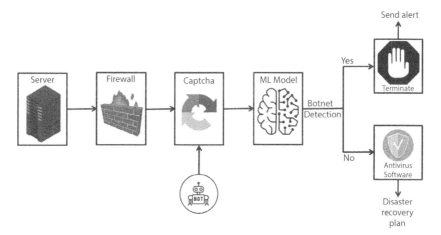

Figure 1.2 System architecture [11].

1.5.2 Future Scope

While we are embracing new ways of digital interaction and more of our critical infrastructure is going digital, the parameters of the transformation underway are not understood by most of us. A better understanding of the global cyberspace architecture is required.

1.6 Conclusion

AI finds its applications in almost every field of science and engineering. AI models need precise safeguards in digital security and new technologies to battle antagonistic machine learning, retain confidentiality, and secure organized learning, and so on. In this chapter, the authors examined specific approaches in AI that are promising and proposed a system of preventing certain types of cyber-security attacks.

References

1. Alex Roney Mathew, Aayad Al Hajj, Khalil Al Ruqeishi (2010), Cyber Crimes: Threats and Protection, *International Conference on Networking and Information Technology, Manila.*
2. Cerli and D. Ramamoorthy (2015), Intrusion Detection System by Combining Fuzzy Logic with Genetic Algorithm, *Global Journal of Pure and Applied Mathematics (GJPAM)*, vol. 11, no. 1.
3. L.S. Wijesinghe, L.N.B. De Silva, G.T.A. Abhayaratne, P. Krithika, S.M.D.R. Priyashan, DhishanDhammearatchi (2016), Combating Cyber Using Artificial Intelligence System, *International Journal of Scientific and Research Publications*, vol. 6, no. 4.
4. Naveen Kumar, Prakarti Triwedi, Pramod Singh Rathore, "An Adaptive Approach for image adaptive watermarking using Elliptical curve cryptography (ECC)", *First International Conference on Information Technology and Knowledge Management* pp. 89–92, ISSN 2300-5963 ACSIS, Vol. 14 DOI: 10.15439/2018KM19
5. Pramod Singh Rathore, "An adaptive method for Edge Preserving Denoising," International Conference on Communication and Electronics Systems, Institute of Electrical and Electronics Engineers & PPG Institute of Technology (2017). *Proceedings of the 2nd International Conference on Communication and Electronics Systems (ICCES 2017)*: 19-20 October, 2017.
6. R. Hill (2010), Dealing with cyber security threats: International cooperation, ITU, and WCIT, *7th International Conference on Cyber Conflict: Architectures in Cyberspace.*
7. Onashoga, S. Adebukola, Ajayi, O. Bamidele and A. Taofik (2013), "A Simulated Multiagent-Based Architecture for Intrusion Detection System", (IJARAI) *International Journal of Advanced Research in Artificial Intelligence*, vol. 2, no. 4.
8. S. Dilek, H. Çakır and M. Aydın (2015), Application of Artificial Intelligence Techniques to Combating Cyber Crimes: A Review, *International Journal of Artificial Intelligence & Applications (IJAIA)*, vol. 6, no. 1.
9. S. Singh and S. Silakari (2009), A Survey of Cyber Attack Detection Systems, *IJCSNS International Journal of Computer Science and Network Security*, vol. 9, no. 5
10. Singh Rathore, P., Kumar, A., & Gracia-Diaz, V. (2020). A Holistic Methodology for Improved RFID Network Lifetime by Advanced Cluster

Head Selection using Dragonfly *Algorithm. International Journal of Interactive Multimedia and Artificial Intelligence,* 6 (Regular Issue), 8. http://doi.org/10.9781/ijimai.2020.05.003

11. Dr. Ritu Bhargava, Pramod Singh Rathore, Rameshwar Sangwa, February 18 Volume 4 Issue 2, "An Contemplated Approach for Criminality Data using Mining Algorithm", *International Journal on Future Revolution in Computer Science & Communication Engineering (IJFRSCE),* pp. 236–240.

2

Privacy Preserving Using Data Mining

Chitra Jalota* and Dr. Rashmi Agrawal

Manav Rachna International Institute of Research and Studies, Faridabad, India

Abstract

On the one hand, data mining techniques are useful to extract hidden knowledge from a large pool of data but on the other hand a number of privacy threats can be introduced by these techniques. The main aim of this chapter is to discuss a few of these issues along with a comprehensive discussion on various data mining techniques and their applications for providing security. An effective classification technique is helpful to categorize the users as normal users or criminals on the basis of the actions which they perform on social networks. It guides users to distinguish among a normal website and a phishing website. It is the task of a classification technique to always alert users from implementing malicious codes by labelling them as malicious. Intrusion detection is the most important application of data mining by applying different data mining techniques to detect it effectively and report the same in actual time so that essential and required arrangements can be made to stop the efforts made by the trespasser.

Keywords: Data mining, security, intrusion detection, anamoly detection, outlier detection, classification, privacy preserving data mining

2.1 Introduction

A computer system has the ability to protect its valuable information, raw data along with its resources in terms of privacy, veracity and authenticity; this ability is known as computer security. A third party cannot read or edit the contents of a database by using the parameters i.e., Privacy/confidentiality and integrity. By using the parameter authenticity, an unauthorised person is not allowed to modify, use or view the contents of a

*Corresponding author: Chitra19878@gmail.com

Neeraj Bhargava, Ritu Bhargava, Pramod Singh Rathore, and Rashmi Agrawal (eds.) *Artificial Intelligence and Data Mining Approaches in Security Frameworks*, (11–32) © 2021 Scrivener Publishing LLC

database. When one or more resources of a computer compromises the availability, integrity or confidentiality by an action, it is known as intrusion. These types of attacks can be prevented by using firewall and filtering router policies. Intrusions can happen even in the most secure systems and therefore it is advisable to detect the same in the beginning. By employing data mining techniques, patterns of features of a system can be detected by an intrusion detection system (IDS) so that anomalies can be detected with the help of an appropriate set of classifiers. For easy detection of intrusion, some important data mining techniques such as classification and clustering are helpful.

Test data could be analysed and labelled into known type of classes with the help of classification techniques. For objects grouping into a set of clusters, clustering methods are used. These methods are used in such a way that a cluster has all similar objects. There could be some security challenges for mining of underlying knowledge from large volumes of data as well as extraction of hidden patterns by using data mining techniques (Ardenas *et al.*, 2014). To solve this issue, Privacy Preserving Data Mining (PPDM) is used, which aims to derive important and useful information from an unwanted or informal database (Friedman, Schuster, 2008). There are various PPDM approaches. On the basis of enforcing privacy principle, some of them can be shown in Figure 2.1.

a) Suppression

An individual's private or sensitive information like name, salary, address and age, if suppressed prior to any calculation is known as suppression. Suppression can be done with the help of some techniques like Rounding (Rs/- 15365.87 can be round off to 15,000), Full form (Name Chitra Mehra can be substituted with the initials, i.e., CM and Place India may be replaced with IND and so on). When there is a requirement of full access to sensitive values, suppression cannot be used by data mining. Another way to do suppression is to limit rather than suppress the record's sensitive information. The method by which we can suppress the identity linkage of a record is termed as De-identification. One such de-identification technique is k-Anonymity. Assurance of protection of data which was released against re-identification of the person's de-identification (Rathore *et al.*, 2020), (Singh, Singh, 2013). K-anonymity and its application is difficult before collecting complete data at one trusted place. For its solution, secret sharing technique based cryptographic solution could be used.

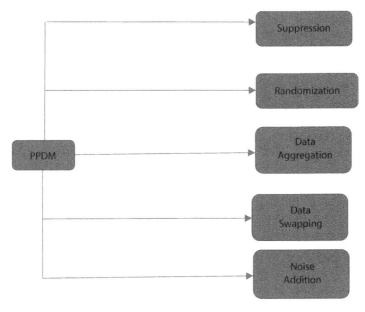

Figure 2.1 Privacy preserving data mining approaches.

b) Data Randomization

The central server of an organization takes information of many customers and builds an aggregate model by performing various data mining techniques. It permits the customers to present precise noise or arbitrarily bother the records and to find out accurate information from that pool of data. There are several ways for introduction of noise, i.e., addition or multiplication of the randomly generated values. To achieve preservation of the required privacy, we use agitation in data randomization technique. To generate an individual record, randomly generated noise can be added to the innovative data. The noise added to the original data is non-recoverable and thus leads to the desired privacy.

Following are the steps of the randomization technique:

1. After randomizing the data by the data provider, it is to be conveyed to the Data Receiver.
2. By using algorithm of distribution reconstruction, data receiver is able to perform computation of distribution on the same data.

c) Data Aggregation

Data is combined from various sources to facilitate data analysis by data aggregation technique. By doing this, an attacker is able to infer private- and individual-level data and also to recognize the resource. When extracted data allows the data miner to identify specific individuals, privacy of data miner is considered to be under a serious threat. When data is anonymized immediately after the aggregation process, it can be prevented from being identified, although, the anonymized data sets comprise sufficient information which is required for individual's identification (Kumar *et al.*, 2018).

d) Data Swapping

For the sake of privacy protection, exchange of values across different records can be done by using this process. Privacy of data can still be preserved by allowing aggregate computations to be achieved exactly as it was done before, i.e., without upsetting the lower order totals of the data. K-anonymity can be used in combination with this technique as well as with other outlines to violate the privacy definitions of that model.

e) Noise Addition/Perturbation

For maximum accuracy of queries and diminish the identification chances its records, there is a mechanism provided by addition of controlled noise (Bhargava *et al.*, 2017). Following are some of the techniques used for noise addition:

1. Parallel Composition
2. Laplace Mechanism
3. Sequential Composition

2.2   Data Mining Techniques and Their Role in Classification and Detection

Malware computer programs that repeat themselves for spreading out from one computer to another computer are called worms. Malware comprises adware, worms, Trojan horse, computer viruses, spyware, key loggers, http worm, UDP worm and port scan worm, and remote to local worm, other malicious code and user to root worm (Herzberg, Gbara, 2004). There are various reasons that attackers write these programs, such as:

i)   Computer process and its interruption
ii) Assembling of sensitive information
iii) A private system can gain entry

It is very important to detect a worm on the internet because of the following two reasons:

i) It creates vulnerable points
ii) Performance of the system can be reduced

Therefore, it is important to notice the worm ot the onset and categorize it with the help of data mining classification algorithms. Given below are the classification algorithms that can be used; Bayesian network, Random Forest, Decision Tree, etc. (Rathore *et al.*, 2013). An underlying principle is that the intrusion detection system (IDS) can be used by the majority of worm detection techniques. It is very difficult to predict that what will be the next form taken by a worm.

There is a challenge in automatic detection of a worm in the system. Intrusion Detection Systems can be broadly classified into two types:

i) On the basis of network: network packets are reflected till that time unless they are not spread to an end-host
ii) On the basis of host: Those network packets are reflected which have already been spread to the end-host

Furthermore, encode network packets is the core area of host-based detection IDS to hit the stroke of the internet worm. We have to pay attention towards the performances of traffic in the network by focusing on the without encoding network packet. Numerous machine learning techniques have been used for worm and intrusion detection systems. Thus, data mining and machine learning techniques are essential as well as they have an important role in a worm detection system. Numerous Intrusion Detection models have been proposed by using various data mining schemes. To study irregular and usual outlines from the training set, Decision Trees and Genetic Algorithms of Machine Learning can be employed and then on the basis of generated classifiers, there could be labeled as Normal or Abnormal classes for test data. The labelled data, "Abnormal", is helpful to point out the presence of an intrusion.

a) Decision Trees

One of the most popular machine learning techniques is Quinlan's decision tree technique. A number of decisions and leaf nodes are required to construct the tree by following divide-and conquer technique (Rathore *et al.*, 2013). A condition needs to be tested by using attributes of input data with the help of each decision node to handle separate outcome of the test. In decision tree, we have a number of branches. A leaf node is represented by the result of decision. A training data set T is having a set of n-classes {C1, C2,..., Cn} when the training dataset T comprises cases belonging to a single class, it is treated as a leaf. T can also be treated as a leaf if T is empty with no cases. The number of test outcomes can be denoted by k if there are k subsets of T i.e. {T1, T2, ..., Tk}, where. The process is recurrent over each Tj, where $1 <= j <= n$, until every subset does not belong to a single class. While constructing the decision tree, choose the best attribute for each decision node. Gain Ratio criteria are adopted by the C4.5 Decision Tree. By using this criterion, the attribute that provides maximum information gain by decreasing the bias/favoritism test is chosen. Thus, to classify the test data that built tree is used whose features and features of training data are the same. Approval of the above test can be done by starting from the root node. On the basis of the result, a branch that leads to a child must be followed. The process would be repeated recursively for the time until the child is not a leaf. To examine a class and its corresponding leaf, test cases must be applied.

b) Genetic Algorithms (GA)

It is used to solve a problem by using biological evolution techniques with the help of machine learning approach. A population of candidate solutions can be optimized with the help of Genetic Algorithm. In genetic algorithm genetic operators, i.e., selection, crossover and mutation are helpful for data structures modelling on chromosomes (Fu *et al.*, 2006). In the beginning, random generation of a population of chromosomes could be performed. In this way, there will be all possible solutions of a problem in the population and that is considered as the candidate solutions. Dissimilar locations of a chromosome called "genes" which can be determined as numbers, characters or bits. To evaluate the goodness of each chromosome on the basis of the desired solution, we use fitness function. Natural reproduction can be stimulated by crossover operator whereas mutation of the species is stimulated by mutation operator. Fittest chromosomes can be chosen by the selection operator (Manek *et al.*, 2016). Genetic Algorithms and its operations can be represented by Figure 2.2. Following are three important factors which we have to consider before using genetic algorithm for solving various problems.

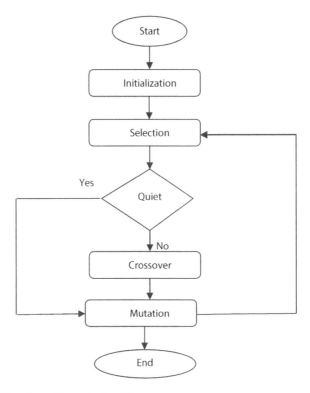

Figure 2.2 Flowchart of genetic algorithm.

1. Fitness function
2. Individuals representation
3. Genetic algorithms parameters

For designing an artificial immune system, genetic algorithm-based method can be used. By using this method, a method for smartphone malware detection has been proposed by Bin *et al.* (Wu *et al.*, 2015). In this approach, static and dynamic signatures of malwares were extracted to obtain the malicious scores of tested samples.

c) Random Forest
It is a classification algorithm that uses collection of tree structured classifiers. In this algorithm, a class is chosen as winner class on the basis of votes given by an individual tree of the forest. To construct a tree, there is a requirement of arbitrary data from a training dataset. Thus, the selected dataset could be divided into training dataset and test dataset. Training data comprises the major portion of the dataset whereas the test data will

have the minor portion of the dataset. Following are the steps required for the tree construction:

1. A sample of N cases is arbitrarily selected from the original dataset which represents the training set required for growing the tree.
2. Out of the M input variables, m variables can be selected arbitrarily. Value of m will be constant at the time of growing the forest.
3. Maximum possible value can be given to each tree in the forest. There is no requirement of trimming or Pruning of the tree.
4. To form the random forest, all classification trees can be combined. The problem of overfitting on large dataset can be fixed with the help of random forest. It can also train/ test quickly on complex data set. It can also be referred as Operational Data mining technique.

Each and every classification tree can be used to cast vote for a class because of its special feature. On the basis of maximum votes assigned to a class, a solution class is built.

d) Association-rule mining

It is used to find fascinating relationships among a set of attributes in datasets (Dwork *et al.*, 2006). Association rule can be defined as inter-relationship of a dataset. It is very helpful to build strategic decisions about different actions like shelf management, promotional pricing, and many more (Jackson *et al.*, 2007). Earlier, a data analyst was involved in associa-tion rule mining whose task is to discover patterns or association rules in the dataset given to him (Rathore, 2017). It is possible to attain sophisti-cated analysis on these extremely large datasets in a cost-effective manner (Tseng *et al.*, 2016), but there may be a chance of data security risk (Beaver *et al.*, 2009) for the data possessor because data miner cans mines sen-sitive information (Bhargava *et al.*, 2017). Nowadays, in knowledge data discovery (KDD) association rule mining is extensively used for pattern discovery. A problem of (ARM) can be solved by navigating the items in a database with the help of various algorithms on the basis of user's require-ment (Patel *et al.*, 2014). Association rule mining (ARM) algorithms can be broadly classified into DFS (Depth First Search) and BFS (Breadth First Search) on the basis of approach used for traversing the search space (Stanley, 2013). These two methods, i.e., DFS (Depth First Search) and BFS

(Breadth First Search) are further divided into methods – intersecting and counting, on the basis of item sets and their support value. The algorithms Apriori-DIC, Apriori and Apriori-TID are BFS-based counting strategies algorithms, whereas partition algorithms are intersecting strategies BFS algorithms. The Equivalence Class Clustering and bottom-up Lattice Traversal (ECLAT) algorithm works on the intersecting strategy with DFS. DFS with Counting strategies comprises FP-Growth algorithm (Yeung, Ding, 2003), (Bloedorn *et al.*, 2003). For improvement in speed, these algorithms can be optimized specifically (Barrantes *et al.*, 2001), (Reddy *et al.*, 2011).

Breadth First Search (BFS) with Counting Occurrences: An eminent algorithm in this group is Apriori algorithm. By clipping the candidates with rare subsets and with the help of this algorithm, the downward closure property of an itemset can be utilized. It should be done before counting their support. Two important parameters to be measured at the time of association rule evaluation which is: support and confidence. In BFS, it is possible to do desired optimization by knowing the support values of all subsets of the candidates in advance. The main drawback of the above mentioned is the increment in computational complexity in a rule that has been extracted from a large database. An improved, dispersed and unsecured form of the Apriori algorithm is Fast Distributed Mining (FDM) algorithm (Lee *et al.*, 1999). Organizations are able to use data more competently with the help of advancements in data mining techniques.

It is possible in Apriori to count the candidates of a cardinality k with the help of a single scan of a large database. Most important limitation of apriori algorithm is to look up the candidates in each transaction. To do the same, a hash tree structure is used (Jacobsan *et al.*, 2014). An extension of Apriori, i.e., Apriori-TID, signifies the current candidate on which each transaction is based, while a raw database is sufficient for a normal Apriori. Apriori and Apriori-TID when combined form Apriori-Hybrid. A prefix-tree is used to fix up the parting that occurs between the processes, counting and candidate generation in Apriori-DIC.

2.3   Clustering

A data mining technique is used for grouping a set of objects in such a way that there is more similarity in the objects of the same class as compared to the objects of the other class. It means cluster of same class, i.e., similarity of intra-cluster is maximum and similarity of inter-cluster is minimum.

Unsupervised learning can be performed with the help of clustering. Following are the types of clustering algorithms:

a) Distribution Based
b) Density Based
c) Centroid Based
d) Connection Based or Hierarchical Clustering
e) Recent Clustering Techniques

a) Distribution-Based Clustering
A model of clustering in which the date is grouped/fitted in the model on the basis of probability, i.e., in what way it may fit into the same distribution. Thus, the groups formed will be on the basis of either normal distribution or *gaussian* distribution

b) Density-Based Clustering
In this type of clustering, a cluster is formed with the help of area with higher density as compared to the rest of the data.

Following are three most frequently used Density-based Clustering techniques:

i) Mean-Shift
ii) OPTICS
iii) DBSCAN

c) Centroid-Based Clustering
Clusters that are represented by a vector are a part of centroid-based clustering. It is not a mandate requirement that these clusters should be a part of the given dataset. The number of clusters is inadequate to size k in k means clustering algorithm; therefore, it is essential to find centres of k cluster and allocate objects to their nearest centres. By taking different values of k random initializations, this algorithm runs multiple times to select the best of multiple runs (Giannotti *et al.*, 2013). In k medoid clustering, clusters are firmly limited to the members of the dataset, whereas in k medians clustering, median is taken to form a cluster; the foremost drawback of these techniques is that we have to select the number of clusters beforehand.

d) Connection-Based (Hierarchical) clustering
As the name itself suggests, this type of clustering is performed on the basis of closeness or distance of objects. The most important key point to form

these types of cluster is the distance between the objects by which they can be connected with each other and form clusters. Instead of single partitioning of dataset, these algorithms provide an in-depth hierarchy of merging clusters at particular distances. To represent clusters, a dendrogram is used. Merging distance of the clusters is shown on the y-axis and an object placement shows the x-axis to ensure that there should not be the mixing of clusters.

On the basis of the different ways with which distance is calculated, there are several types of connection-based clusters:

    i)   Single-Linkage Clustering
    ii) Complete-Linkage
    iii) Average-Linkage Clustering

e) Recent Clustering Techniques

For high dimensional data, the above-mentioned standard clustering techniques are not fit, therefore some new techniques are being discovered. These new techniques can be classified into two major categories, namely: Subspace Clustering and Correlation Clustering.

A small list of attributes that should be measured for the formation of a cluster is taken into consideration under subspace clustering. Correlation between the chosen attributes can also be performed with correlation clustering.

2.4   Privacy Preserving Data Mining (PPDM)

To extract the pertinent knowledge from large volumes of data and to protect all sensitive information of that database, we use privacy preserving data mining (PPDM). These techniques are created with the aim to confirm the protection of sensitive data so that privacy can be reserved with the efficient performance of all data mining operations. There are two classes of privacy concerned data mining techniques:

    1. Data privacy
    2. Information privacy
        Modification of database for the protection of sensitive data of the individuals, we use data privacy technique. If there is a requirement for the modification of sensitive knowledge that can be deduced from the database, information privacy technique is preferred. To provide privacy

to input, data privacy is preferable, whereas for providing privacy to output, the technique of information privacy is used. To reserve personal information from exposure is the main focus of a PPDM algorithm. It relies on the analysis of those mining algorithms that are attained during data privacy. Main objective of Privacy Preserving Data Mining is building algorithms that convert the original data in some useful means, so that there is no visibility of private data and knowledge even after a successful mining process. Privacy laws would allow the access in the case that some related satisfactory benefit is found resulting from the access.

2.5   Intrusion Detection Systems (IDS)

Onset detection of the intrusion is the main aim of an Intrusion detection system. There is a requirement of a high level of human knowledge and substantial amount of time to attain security in data mining. However, intrusion detection systems based on data mining need less expertise for better performance. To perceive network attacks in contrast to services that are vulnerable, intrusion detection system is very helpful. Applications-based data-driven attacks always privilege escalation (Thabtah *et al.*, 2005), un-authorized logins and files accessibility is very sensitive in nature (Hong, 2012). Data mining process can be used as a tool for cyber security for the competent detection of malware from the code. Figure 2.3 shows the outline of an intrusion detection system. Several components such as, sensors, a console monitor and a central engine forms the complete intrusion detection system. Security events are generated by sensors whereas the task of console monitor is to monitor and control all events and alerts. The main function of the central engine is recording of events in a database and on the basis of these events, alerts can be created followed by certain set of rules. Following factors are responsible for the classification of an intrusion detection system:

i) Location
ii) Type of Sensors
iii) Technique used by the Central engine for generation of alerts.

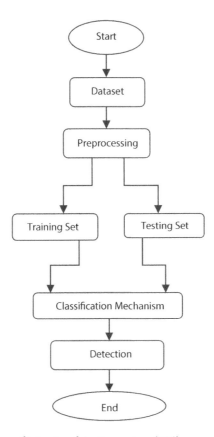

Figure 2.3 An overview of intrusion detection system (IDS).

All the three components of an intrusion detection system can be integrated into a single device.

2.5.1   Types of IDS

Detection of an intrusion could be done either on a network or with an individual system and therefore we have three types of IDS, namely: Network Based, Host Based and Hybrid IDS.

2.5.1.1   Network-Based IDS

Computer networks have been targeted by enemies and criminals because of their progressively dynamic roles in modern societies. It is very important to find the best possible solutions for the sake of protection of our systems. Various techniques of intrusion prevention like programming errors avoidance, protection of information using encryption techniques and biometrics or passwords (Zhan *et al.*, 2005) can be used as a first line of security. By using intrusion prevention technique as the only protection measure, our system is not 100% safe from combat attacks. To provide an additional security for computer system, the above-mentioned techniques are used. Various resources like accounts of users, their file systems and the system kernels of a target system must be protected by an intrusion detection system. For network-based intrusion detection systems, data source is the network packets. To listen and analyse network traffic as the packets travel across the network, the network-based intrusion detection system (NIDS) makes use of a network adapter. A network-based intrusion detection system is used to generate alerts for the detection of an intrusion which is outside of the boundary of its enterprise.

Advantages
Following are the advantages of a Network-Based IDS:

1. They can be made invisible to improve the security against attacks.
2. Large size of networks can be monitored by network-based IDS.
3. This IDS can give better output deprived of upsetting the usual working of a network.
4. It is easy to fit in an IDS into an existing network.

Limitations
Limitations of Network-Based IDS are as follows:

1. Virtual private networks encrypted information cannot be analysed with network-based IDS.
2. Successful implementation of network-based IDS is based on the intermediate switches present in the network.
3. Network-based IDS would be unstable and crash when the attackers splinter their packets and release them.

2.5.1.2   Host-Based IDS

In this type of IDS, various logs can be screened with the help of sensors that are placed on network resources. These logs are generated by the host operating system or application programs. Certain events or actions which may occur at individual network resource are recorded by audit logs. These types of IDS can handle even those attacks that cannot be handled. Because of this, an attacker can misuse one of trusted insiders (Desale *et al.*, 2005). Signature rule base that is derivative from security policy which is specific to a site is utilized by a host-based system. All the problems associated with a Network-based IDS can be overcome by host-based IDS as it can alert the security personnel with the location details of intrusion. Accordingly, the person can take instant action to stop the intrusion.

Advantages
Following are the advantages of Host-Based IDS:

1. It can perceive even those attacks that are not detected by a Network-Based IDS.
2. For the detection of attacks concerning software integrity breaches, it works on audit log trails of operating system.

Disadvantages
Disadvantages of Host-Based IDS are as follows:

1. Various types of DoS (Denial of Service) attacks can disable the Host-Based IDs.
2. Attacks that target the network cannot be detected by host-based IDS.
3. To configure and manage every individual system is very difficult.

2.5.1.3   Hybrid IDS

It is a combination of network and host-based IDS to form a structure for next-generation intrusion detection systems. This arrangement is generally known as a fusion/hybrid intrusion detection system. By adding network based and host-based IDS, it would significantly improve resistance against few more attacks. Data mining techniques required for IDS are Pattern Matching, Classification and Feature Selection Pattern Matching.

2.6   Phishing Website Classification

It is a kind of social engineering attack generally used to filch data of a user, like login credentials and credit card numbers. To cover up honest websites, forged websites are usually formed by fraudulent people. Due to phishing activities of attackers, users mistakenly lose their money. Therefore, a critical step must be taken for the protection of online trading. Goodness of the extracted features denotes the prediction and classification accuracy of a website. An anti-phishing tool is used by most of the internet users to feel safe against phishing attacks. Anti-phishing tool is required to predict accurate phishing. Content parts of phishing websites along with security indicators may have a set of clues within the browsers. Various methods have been proposed to handle the problem of phishing. For predicting phishing attacks, rule-based classification, which is a data mining technique, is used as a proficient method for prediction. If an attacker is sending an email to victims by requesting them to reveal their personal information, it is an indication of phishing. To create phishing websites with proper trick, a set of mutual features are used by phishers. We can distinguish between phishy and non-phishy websites on the basis of extracted features of that visited website.

Identification of phishing sites can be done with the help of two approaches:

  i) Blacklist based: It includes comparative analysis of the URL, i.e., requested along with other URLs which are present in that list.
 ii) Heuristic based: Certain features from various websites are collected and labeled as either as phishy or genuine.

The main drawback of the blacklisted approach is that it cannot have all phishing websites because every second, a new malicious website is launched, while a heuristic-based approach can identify fake websites that are original. Heuristic-based methods depend on the feature's selection and the manner in which they processed. Data mining is used to discover relations and pattern amongst features within a given dataset. The utmost job of data mining is to take decisions because these decisions are patterns and rules dependent which have been derived using the data mining algorithms. Though considerable progress has been made for the development of prevention techniques, still phishing is a threat because the techniques used for countermeasures are still based on blacklisting of reactive URL (Polychronakis, 2009). Because of the shorter lifetime of phishing websites, methods used in these sites are considered as ineffective. A new approach, associative classification (AC) was found more appropriate for these kinds of applications; it is a mixture of Association rule and Classification techniques of data mining.

There are two stages in association classification (AC):

i) Training phase: It is used to induce hidden knowledge (rules) with the help of Association rule.
ii) Classification phase: It is used to build a classifier after cropping ineffective and superfluous rules.

It has been proved from many research studies that association classifier (AC) generally shows better classifiers in terms of error rate than decision tree and rule induction (standard classification approaches).

2.7   Attacks by Mitigating Code Injection

Code injection attack is a technique to write new machine code into the susceptible program's memory. If there is a bug in the program, the control can be sent to the new code after manipulating it. W+X, the protection technique (Diwate, Sahu, 2014) alleviates the code injection attack by permitting one operation, i.e., either to write or execute operations but not both simultaneously (Mitchell, Chen, 2013).

2.7.1   Code Injection and Its Categories

Following are the types of code injection attacks:

i) SQL Injection
ii) HTML Script Injection
iii) Object Injection
iv) Remote File Injection
v) Code Reuse Attacks (CRAs).

i. SQL Injection: It can be defined as a technique by which SQL syntax can be used to input commands for reading, alteration or modification of a database.

For example, there is a field on a web page regarding authentication for user password. Generally, we use script code for this. This script code will generate a SQL query so that matching password entered against the list of user names could be verified:

SELECT User List. Username FROM User List WHERE User List. Password = 'Password'

ii. HTML Script Injection: Malicious code can be injected by an attacker with the help of tags. Thus, location property of the document would be changed by setting it to an injected script.

iii. Object Injection: Hypertext pre-processor (PHP) is used for serialization and deserialization of objects. With the help of object injection, existing classes in the program can be modified and malicious attacks can be executed if an untrustworthy input is allowed into the deserialization function.

iv. Remote File Injection: To cause the intended destruction, remote infected file name could be provided by attackers by alter the path command of the script file as the path.

v. Code Reuse Attacks: Code reuse attacks (CRAs) are recent development in security. They occur when an attacker expresses the flow of control through a previously existing code. By using this, attackers are allowed to execute random code on a compromised machine. These are return-oriented and jump-oriented programming approaches. They can reclaim library code fragments. The Return Into Lib C (RILC) is a type of code-reuse attack where the stack is compromised and the control is transferred to the beginning of an existing library function such as mprotect() to create a memory region that allows both write and execution operations on it to bypass W+X (Bhatkar *et al.*, 2005). To overcome such attacks, we use data mining techniques. When the source code is checked to reveal any such fault and for this the instructions are classified as malicious. Some of the classification algorithms that can be used in this Regard are Logistic Regression, Bayesian, Support Vector Machine and Decision Tree.

2.8   Conclusion

The main aim of this study is to find the role of Data Mining techniques in attaining security. A few applications such as Privacy Preserving Data Mining (PPDM), Intrusion Detection System (IDS), Phishing Website Classification and Mitigation of Code Injection are discussed. Some Classification and Clustering algorithms are also discussed for their significant role in an intrusion detection system. Other basic Data mining techniques used for intrusion detection system such as Feature Extraction, Association Rule Mining and Decision Trees are also discussed. Other security applications of Data Mining such as Malware Detection, Spam Detection, Web Mining and Crime Profiling can also be explored in terms of security as a future scope.

References

Cárdenas, A. A., Berthier, R., Bobba, R.B., Huh, J.H., Jetcheva, J.G., Grochocki, D., & Sanders, W.H. (2014) "A Framework for Evaluating Intrusion Detection Architectures in Advanced Metering Infrastructures," *IEEE Transactions on Smart Grid*, vol. 5(2), pp. 906–915.

Friedman, R. W., & Schuster. A. (2008) "Providing kAnonymity in Data Mining," *VLDB Journal*, vol. 17(4), pp. 789–804.

Singh, R., Kumar, P. & Diaz, V. (2020) "A Holistic Methodology for Improved RFID Network Lifetime by Advanced Cluster Head Selection using Dragonfly Algorithm" *International Journal of Interactive Multimedia and Artificial Intelligence*, vol. 6(2), pp. 8.

Singh, B., Singh, R. & Rathore. P.S. (2013) "Randomized Virtual Scanning Technique for Road Network" *International Journal of Computer Applications*, vol. 77(16). pp. 1-4.

Kumar, N., Triwedi, P. & Rathore, P.S. (2018) "An Adaptive Approach for image adaptive watermarking using Elliptical curve cryptography (ECC)" *First International Conference on Information Technology and Knowledge Management* pp. 89–92, ISSN 2300-5963.

Bhargava, N., Singh, P., Kumar, A., Sharma, T. & Meena, P. (2017) "An Adaptive Approach for Eigenfaces-based Facial Recognition" *International Journal on Future Revolution in Computer Science & Communication Engineering (IJFRSCE)*, vol. 3(12), pp. 213 – 216.

Herzberg, A. & Gbara, A. (2004) "Trustbar: Protecting (even naive) Web Users from Spoofing and Phishing Attacks" *Cryptology ePrint Archive Report* pp. 155.

Rathore, P. S., Chaudhary A. & Singh, B. (2013) "Route planning via facilities in time dependent network," *IEEE Conference on Information & Communication Technologies*, pp. 652-655.

Fu, A. Y,, Wenyin, L. & Deng X (2006) "Detecting Phishing Web Pages with Visual Similarity Assessment Based on Earth Mover's Distance (emd)," *IEEE Transactions on Dependable and Secure Computing*, vol. 3(4), pp. 301–311.

Manek, A., S., Shenoy, P., D., Mohan, M., C. & Venugopal K. R., (2016) "Detection of Fraudulent and Malicious Websites by Analysing User Reviews for Online

Shopping Websites," *International Journal of Knowledge and Web Intelligence*, vol. 5(3), pp. 171–189.

Wu, B., Lu, T., Zheng, K., Zhang, D. & Lin, X. (2015) "Smartphone Malware Detection Model Based on Artificial Immune System," *China Communications*, vol. 11(13), pp. 86–92.

Dwork, C., McSherry, F., Nissim, K. & Smith, A. (2006) "Calibrating Noise to Sensitivity in Private Data Analysis," *Theory of Cryptography Conference*, pp. 265–284.

Jackson, C., Simon, D.R., Tan, D. S. & Barth, A. (2007) "An Evaluation of Extended Validation and Picturein-Picture Phishing attacks," *International Conference on Financial Cryptography and Data Security*, pp. 281–293.

Rathore, P.S. (2017) "An adaptive method for Edge Preserving Denoising, International Conference on Communication and Electronics Systems, Institute of Electrical and Electronics Engineers, *Proceedings of the 2nd International Conference on Communication and Electronics Systems* (ICCES 2017).

Tseng, C., Y., Balasubramanyam, P., Limprasittiporn, R., Rowe, J. & Levitt, K. (2016) "A Specification-Based Intrusion Detection System" Global Journals Inc. (US) *Global Journal of Computer Science and Technology*, vol. 16(5), pp.125–134.

Beaver, D., Micali, S. & Rogaway, P. (1990) "The Round Complexity of Secure Protocols," *Proceedings of the 22nd Annual ACM Symposium on Theory of Computing*, pp. 503–513.

Bhargava, N., Dayma, S., Kumar, A. & Singh, P. (2017) "An approach for classification using simple CART algorithm in WEKA," *11th International Conference on Intelligent Systems and Control (ISCO)*, pp. 212–216.

Patel, D., K., B., & Bhatt, S. H. (2014) "Implementnig Data Mining for Detection of Malware from Code," *International Journal of Advanced Computer Technology: Compusoft*, vol. 3(4), pp. 732–740.

Stanley, D. M. (2013) "CERIAS Tech Report 2013-19 Improved Kernel Security through Code Validation, Diversification, and Minimization," Ph.D. Thesis.

Yeung D. Y. & Ding, Y. (2003) "Host-Based Intrusion Detection Using Dynamic and Static Behavioral Models," *Pattern Recognition*, vol. 36(1), pp. 229–243.

Bloedorn, E., Christiansen, A. D., Hill, W., Skorupka, C., Talbot, L.M. & Tivel, J. (2001) "Data Mining for Network Intrusion Detection: How to Get Started," *MITRE*, pp. 1–9.

Barrantes, E.G., Ackley, D. H., Palmer, T.S., Stefanovic, D. & Zovi, D.D. (2003) "Randomized Instruction Set Emulation to Disrupt Binary Code Injection Attacks," *Proceedings of the 10th ACM Conference on Computer and Communications Security*, pp. 281–289.

Reddy, G., Iaeng, M., Reddy, V. & Rajulu (2011) "A Study of Intrusion Detection in Data Mining" *World Congress on Engineering (WCE)*, pp. 6–8.

Lee, W., Stolfo, S.J. & Mok, K.W. (1999) "A Data Mining Framework for Building Intrusion Detection Models," *Proceedings of the IEEE Symposium on Security and Privacy*, pp. 120–132.

Jacobson, E. R., Bernat, A.R., Williams, W.R. & Miller, B.P. (2014) "Detecting Code Reuse Attacks with a Model of Conformant Program Execution," *International Symposium on Engineering Secure Software and Systems*, pp. 1–18.

Giannotti, F., Lakshmanan, L.V., Monreale, A., Pedreschi, D. & Wang, H. (2013) "Privacy-Preserving Mining of Association Rules from Outsourced Transaction Databases," *IEEE Systems Journal*, vol. 7(3), pp. 385–395.

Thabtah, F., Cowling, P., & Peng, Y. (2005) "MCAR: Multiclass Classification based on Association Rule," *3rd ACS/IEEE International Conference on Computer Systems and Applications*, pp. 33–39.

Habibi, J., Panicker, A., Gupta, A. & Bertino, E. (2015) "DISARM: Mitigating Buffer Overflow Attacks on Embedded Devices," *International Conference on Network and System Security*, pp. 112–129.

Zhan, J., Matwin, S. & Chang L (2005), "Privacy Preserving Collaborative Association Rule Mining," *IFIP Annual Conference on Data and Applications Security and Privacy*, pp. 153–165.

Desale, K.S. & Ade, R. (2015) "Genetic Algorithm Based Feature Selection Approach for Effective Intrusion Detection System," *International Conference on Computer Communication and Informatics (ICCCI)*, pp. 1–6.

Polychronakis, M. (2009) "Generic Detection of Code Injection Attacks using Network-Level Emulation," Ph.D. Thesis.

Diwate, & Sahu, A., (2014) "Efficient Data Mining in SAMS through Association Rule," *International Journal of Electronics Communication and Computer Engineering*, vol. 5(3), pp. 593–597.

Mitchell R. & Chen, R. (2013) "Effect of Intrusion Detection and Response on Reliability of Cyber Physical Systems," *IEEE Transactions on Reliability*, vol. 62(1), pp. 199–210.

Bhatkar, S., DuVarney, D. C. & Sekar, R. (2005) "Efficient Techniques for Comprehensive Protection from Memory Error Exploits" *Proceedings of the 14th USENIX Security Symposium*.

3

Role of Artificial Intelligence in Cyber Security and Security Framework

Shweta Sharma

M.Tech Scholar, MDS University, Ajmer, India

Abstract

Artificial Intelligence (AI) is a popular expression in the digital world. It is as yet an emerging science in various features as indicated by the difficulties experienced in the 21st century. Nowadays one can't imagine a world without AI as it has had a gigantic impact on human life. The essential objective of artificial intelligence is to develop advancement-based activities which addresses the human data in order to handle issues. Basically artificial intelligence is an examination of how an individual thinks, works, learns and makes choices in any circumstance of life, whether or not it may be related to basic reasoning or learning new things or thinking equitably or to produce an answer, etc. Computer-based intelligence is in almost every sphere of human life, including gaming, language preparation, discourse acknowledgment, insight robots, money-related exchanges, and so forth; every movement of human life has become a subset of AI. Security issues have become a significant threat for governments, banks, and associations due to online ambushes by software engineers. AI and cyber security have expanded and become more essential in the progressing events but AI is suffering also as it is a dynamic and fragile issue associated with human life.

Keywords: Artificial intelligence, cyber security, role of AI for cyber security, impacts of AI on cyber security, AI security threats, improve cyber security for artificial intelligence

Email: sharmashweta671@gmail.com

Neeraj Bhargava, Ritu Bhargava, Pramod Singh Rathore, and Rashmi Agrawal (eds.) *Artificial Intelligence and Data Mining Approaches in Security Frameworks*, (33–64) © 2021 Scrivener Publishing LLC

3.1 Introduction

Artificial Intelligence

Artificial Intelligence systems can be used to make sense of bothersome data and to engage security professionals to appreciate the advanced condition so as to analyse irregular action. Artificial intelligence can likewise benefit cyber security by creating robotized strategies at whatever point digital dangers are recognized. AI brainpower can dissect huge measures of information and permit the improvement of existing frameworks and programming in a suitable manner to diminish digital assaults. Essentially, the usage of AI for digital security arrangements will assist with shielding associations from existing digital dangers and distinguish new kinds of malware. The solidification of artificial intelligence into security structures can be used to reduce the routinely extending risks of advanced security that is being looked at by the overall associations. By the ventures applications using Machine learning similarly as electronic thinking (AI) are thoroughly being used considerably more as data grouping, storing limits and handling power are extending. Continuously, the enormous size of data is difficult to be dealt with by individuals. With the help of artificial intelligence, the enormous amount of data can in all probability be reduced down in milliseconds, in light of which the endeavour can without a very remarkable stretch recognize and furthermore recover from peril. Clearly protection against astute mechanized weapons can be developed just by wise programming, and events of the most recent two years have indicated quickly developing data on malware and electronic weapons. With the progression in innovation, digital wrongdoings are likewise expanding and getting perplexing [1]. Digital crooks are propelling advanced assaults that are putting present-day security frameworks in danger. Thus, the cyber security business is also advancing to satisfy the expanding security needs of organizations. Be that as it may, these defensive techniques of preservation may likewise come up short sooner or later. To up their game and upgrade their weakness identification components, organizations are picking Artificial Intelligence (AI). AI consciousness in Cyber Security is supporting organizations to defending their protection systems. It is additionally helping them in breaking down digital violations better. Artificial knowledge in digital security is gainful considering the way that it improves how security authorities explore, study, and deal with cybercrime. It improves the advanced security innovations that associations use to fight cybercriminals and helps monitor associations and

Figure 3.1 Artificial intelligence.

clients. Then again, AI can be exceptionally resource serious. Artificial Intelligence reasoning is a developing territory of premium and venture inside the digital security network [2].

Above Figure 3.1 represent the Artificial intelligence for the CyberSecurity where all the factor concerns with cyber are introduced. Computer-based intelligence alludes to advancements that can appreciate, determine, and execute, dependent on obtained and determined data. Artificial intelligence works in three different ways:

1. Helped information, by and large available today, improves what people and organizations are starting at now accomplishment.
2. Augmented knowledge, growing today, enables people and relationship to do things they couldn't previously do.
3. Autonomous insight, being made for the prospect, features machines that follow up by themselves. An instance of this is self-driving vehicles, when they come into broad use.

Albeit still in its outset, AI can be said to have some level of human insight: a store of space explicit information; instruments to secure new information; and components to put that information to use. AI, master frameworks, neural systems, and profound learning are on the whole models or subsets of AI innovation today [3].

- Machine learning utilizes factual approaches to implement PC structures to "learn" (e.g., dynamically improve execution) utilizing information instead of being expressly customized. AI works best when focused on a particular assignment as opposed to a wide-running crucial.

- Expert frameworks are programs intended to take care of issues inside particular spaces. By mirroring the considering human specialists, they tackle issues and settle on choices utilizing fluffy standards-based thinking through cautiously curated assemblages of information.

- Neural systems utilize an organically motivated programming worldview which empowers a PC to gain from observational information. In a neural system, every hub doles out a load to its info speaking to how right or erroneous it is comparative with the activity being performed. The last yield is then controlled by the whole of such loads.

- Deep Learning is a broader gathering of artificial intelligence strategies subject to learning data depictions, as opposed to task-express counts. Today, picture acknowledgment by means of profound learning is regularly superior to people, with an assortment of uses, for example, self-driving vehicles, examine investigations, and clinical determinations.

3.2   AI for Cyber Security

Cyber Security – AI is turning into an incredible instrument for cyber security and infra around. AI functions admirably when its product gets prepared on a huge informational index of cyber security gadgets, organize and any data which is valuable to finish up anything. It targets identifying peculiarity, caution and square. Any strange example is an indication of stress to it. Unfortunately we ought not to overlook that programmers are additionally utilizing AI consciousness in digital assaults that are further developed and harder to distinguish.

Current technologies threaten an association's cyber security. Certainly, even with the new headway in the resistance systems, security proficiency comes up short eventually. Joining the quality of AI with the abilities of security experts from weakness checks to protection turns out to be extremely powerful. Associations get moment bits of knowledge, thusly, get decreased reaction time. AI for Cyber Security is the new wave in Security shown in Figure 3.2 [4].

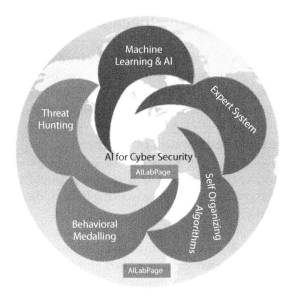

Figure 3.2 AI for cyber security.

➢ Present AI Cyber Security Analytics Solutions for Enterprises:
  - Perspective Analytics: Determination of the activities required for investigation or reaction.
  - Diagnostic Analytics: Evaluation of main driver examination and usual way of doing things of the episodes and assaults.
  - Predictive Analytics: Determination of higher hazard clients and resources later on and the probability of upcoming dangers [4].
  - Detective Analytics: Recognition of covered up, obscure dangers, skirted dangers, progressed malware, and horizontal development.
  - Descriptive Analytics: For acquiring the current status and execution of the measurements and patterns.
  - AI-fuelled Risk Management Approach to Cyber Security:
    - Right Collection of Data.
    - Representation Learning Application.
    - Machine Learning Customization.
    - Cyber Threat Analysis.
    - Model Security Problem

3.3   Uses of Artificial Intelligence in Cyber Security

- In expansion, AI-based digital security frameworks can give successful security norms and help grow better anticipation and recuperation techniques. On the other hand, the utilization of AI for digital security assists with making a dynamic, genuine – time, worldwide verification structure that changes area or system get to benefits.
- Actually, in excess of 90 percent of US and Japanese cyber security experts anticipate that assailants would utilize AI for the organizations they are working for, as indicated by an examination by Webroot.
- Enterprise AI exercises have a wide extent of anticipated shortcomings, including malicious debasement or control of getting ready data, execution and section game plan. Darktrace's cyber security association declares that its modified AI advancement has already perceived 63,500 dark threats on more than 5,000 frameworks, including zero – day abuses, inside risks and unnoticeable, covered attacks [5].
- From framework and web application security to peril protection and united and secure access, Fortune's advanced security things are used by most Fortune 500 associations.
- For associations and organizations that need digital security arrangements, Spark Cognition gives AI items that distinguish and ensure against malware, emancipate, Trojan horse and different dangers.
- Protector orchestrate security show gives information within the threats of a business cloud or crossbreed position.
- The bottom line says that the instrument can quickly recognize uncommon traffic in the framework including bitcoin mining, inaccessible record execution, and even brute force logins – to ensure the security of the entire association.
- Today, it speaks to the most recent innovation for the pragmatic utilization of AI to cyber security.
- Cyber security items gather immense measures of information – the cyber security investigator truly suffocates the information.
- AI offers an immense potential for assisting with defeating the test and advance to assist associations with improving their cyber security mentalities through savvy code investigation and setup examination and movement checking.

- Although AI is utilized in numerous zones, cyber security is one of them that has gotten exceptional consideration in light of the rate at which dangers are being created and the quantity of assaults.
- Many security administrators said they are currently "absolutely needy" on AI innovation to protect their systems and delicate information.
- However, air-based digital security arrangements can identify examples of pernicious conduct in organize traffic and records and sites that are acquainted with the system.
- Because AI consciousness–based system security arrangements don't depend on marks, they can distinguish assaults that are impractical.
- The federal stage is generally coordinated into a bank or business framework and can caution human extortion and hazard examination on the off chance that they are truly viewed as high-chance (dependent on pre-characterized factors), consequently accelerating misrepresentation identification procedures and decreasing bogus constructive outcomes.
- The organization asserts that its foundation can bolster the security and operational exercises of organizations through programmed learning model acknowledgment in chronicled arrange information. Barrier storm says that their SaaS arrangements could furnish IT safety faculty at manages an account with access to the occasion-related information in one spot through a solitary dashboard.
- DefenceStorm says it has coordinated its SaaS investigation answer for update the current Live Oak Bank information and examination frameworks in a couple of months.
- Banks ought to likewise know that such endeavours by AI are essentially planned for gathering and sorting out information, hence guaranteeing that information identified with security, for example, IP addresses, firewall information, and interruption counteraction frameworks, are gathered in a comparable organization.
- Emery's Artificial Intelligence Research enables organizations and supervisors to endure and create AI problematic issues through in – profundity AI Research, guidance and bits of knowledge.

- Cyber security alludes to innovation and practices planned for shielding systems and data from harm or unapproved access.
- Digital security is crucial in light of the fact that law-making bodies, organizations and military forces accumulate methodology and preserve a lot of information on PCs.
- Cyber assailants are placing cash into robotization to dispatch assaults, while numerous associations are as yet investigating the manual exertion to join inward security results and put them in setting with data about outside dangers.
- Most cyber security arrangements utilize a standard-based or signature approach that requires an excess of human mediation and institutional information.
- Artificial knowledge may expand the efficiency of people so as to build the time spent on cyber security.
- So far, the government has to a great extent combined its cyber security frameworks, which has prompted a divided way to deal with security frameworks.
- The utilization of AI and AI neural systems has permitted engineers to adjust to new assault vectors and better foresee the subsequent stages of cybercriminals.
- The utilization of AI could additionally animate comparative assaults and lead to another period of state-supported assaults and digital reconnaissance.
- As an ever-increasing number of organizations are embracing AI-based and AI items as a component of their barrier procedure, specialists are worried this could prompt an incorrect conviction that all is well with the world for representatives and IT experts.

3.4   The Role of AI in Cyber Security

AI reasoning (Artificial intelligence) can be characterized as counterfeit dynamic comparable or equivalent to human dynamic, in light of certain extraordinary calculations and related scientific computations. On the opposite side, Cyber Security refers to the safety efforts to be taken to deal with digital assaults in the virtual world shown in Figure 3.3 [6].

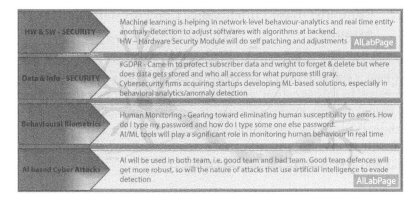

Figure 3.3 Role of artificial intelligence in cyber security.

For cyber security, Artificial intelligence can break down immense measures of information, help the correct frameworks and programming are to settle on the choice and acquire tremendous decreases in assaults and inconsistencies in a lot quicker way. Since it can work 24/7 without rest, it is superior to human workers. Computer-based intelligence will permit computerized programming testing to discover and destroy bugs before they boat to maintain a strategic distance from any financial open doors on escape clauses.

3.4.1   Simulated Intelligence Can Distinguish Digital Assaults

Simulated intelligence can be effectively used to recognize unforeseen assaults on the internet, including diverse web stages and high-security-based authority sites. Programmers utilize various approaches to start digital assaults and interest for recover. In this situation, destinations needing high security depend upon simulated intelligence as the essential strategy to recognize digital assaults. In addition, it is hard for programmers to gain admittance to high-security sites since high-security sites rely on man-made intelligence to identify such an unapproved section. A high achievement pace of simulated intelligence-based sites in identifying digital assaults leads comparable destinations to pick computer-based intelligence as the essential safety effort.

3.4.2   Computer-Based Intelligence Can Forestall Digital Assaults

One can see that unimportant recognizable proof of a security danger can't help a site of virtual stage to avoid digital aggressors including programmers. In this situation, artificial intelligence can be utilized to forestall digital assaults in various ways. In order to forestall a digital assault, the individual who is responsible for a site must think like a programmer thinks. Here, computer-based intelligence can utilize how a programmer thinks and acts to break the security code.

3.4.3   Artificial Intelligence and Huge Scope Cyber Security

Envision a site with less traffic, a couple of interconnected frameworks and less prominence. There is no utilization to rely on complex artificial intelligence to shield the site from digital assault. In addition, programmers may not target less well-known sites and open stages since they can't increase much from the proposed assault [7].

Developing innovations put cyber security at serious risk. In fact, even the new degrees of progress in defensive frameworks of security specialists miss the mark eventually. Likewise, as threatening defensive systems and advancements are running in an endless cycle, the multifaceted nature and volume of cyber-attacks have extended. Solidifying the nature of Arterial awareness with cyber security, security specialists have additional resources for defend frail frameworks and data from computerized attackers. In the wake of applying this innovation, it brought bits of knowledge, bringing about decreased reaction times. Cap Gemini as of late released a report dependent on simulated intelligence in cyber security, which specifies that 42% of the organizations considered had seen a rise in security episodes through time-sensitive applications. It additionally discovered that two out of three associations want to embrace man-made intelligence arrangements by 2020. Data security is right now a bigger issue than ever before. Reviving existing cyber security game plans and maintaining each possible applicable security layer doesn't ensure that your data is secure. Regardless, having a strong assistance of cutting-edge advancements will encourage the task of security specialists [8].

3.4.4 Challenges and Promises of Artificial Intelligence in Cyber Security

Cyber security isn't only a data innovation division or an issue involving individuals in a similar office. It is the activity of each worker and even clients in the city.

While cyber security specialists have acknowledged artificial intelligence as the eventual fate of the business, discovering answers for its issues are as yet not sufficiently tended to. Aside from being an answer, it is a significant danger to organizations [9].

AI reasoning can viably explore customer rehearses, finish up a model, and perceive a wide scope of varieties from the standard or variations from the norm in the framework. With such data, it's significantly less complex to perceive advanced shortcomings quickly. Then again, the obligations which are by and by dependent on human information will by then be weak to threatening computerized ventures imitating bona fide mimicked insight-based computations. A couple of organizations are hustling into

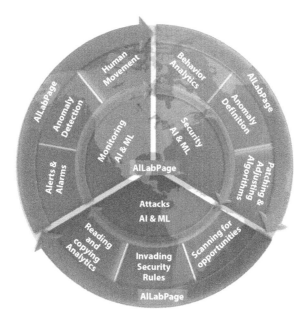

Figure 3.4 Challenges in cyber security.

getting their AI-based things out in the market. With this lead, they may disregard the seriousness of the situation, drawing the wrong conclusion that everything is great with the world. Contingent upon "directed learning" is another threat. Under this, the estimations name the educational assortments as indicated by their demeanour. It could be malware, clean data, or some other tag. Cybercriminals, if they gain access to the security firm, can change the name to their advantage. Also, routine endeavours relying upon PC-based knowledge can be constrained by cutting-edge hacking endeavours utilizing artificial intelligence [6].

Regardless of being a security danger to the associations, computerized reasoning will continue constraining the typical security commitments with first-rate results. Computerized reasoning automation will have the choice to perceive rehashing events and even remediate them. It will in like manner have the alternative to direct insider threats and device the executives describe in above Figure 3.4 [11].

3.4.5   Present-Day Cyber Security and its Future with Simulated Intelligence

Today, businesses and other organizations give close thought to their framework security. They think about the colossal impact of every little to gigantic extension computerized attack. To ensure against such attacks, they use different lines of gatekeeper. This multi-layered security structure generally starts with the best sensible firewall fit for controlling and filtering through the framework traffic. After this layer, the second line of defense includes antivirus programming (AV Programming). These AV instruments check through the system to find and take out vindictive codes and records. With these two lines of shield, organizations reliably run fortifications as a bit of a disaster recovery plan [12].

For the present, setting up firewall strategies, overseeing reinforcements, and numerous such errands require an expert, yet artificial intelligence will change the conventional methodology.

- Organizations will have the option to screen and react to security occurrences by utilizing propelled apparatuses.
- The cutting-edge firewalls will have in-constructed AI innovation that could discover an example in organize bundles and square them naturally whenever hailed as a danger.
- Predictably, the characteristic language abilities of artificial intelligence will be utilized to comprehend the beginning

of digital assaults. This hypothesis can be incorporated by examining information over the web.

3.4.6 Improved Cyber Security with Computer-Based Intelligence and AI (ML)

Convoluted hacking methods, for instance, confusing, polymorphism, and others, make it a certifiable test to perceive harmful activities. Moreover, security engineers with region express workforce lack are another issue. With man-made reasoning wandering into cyber security, authorities and examiners are endeavouring to use its capacity to recognize and check progressed advanced ambushes with irrelevant human intervention. Artificial intelligence reasoning frameworks and AI, a subset of PC-based insight, has engaged security specialists to get some answers concerning new attack vectors [13].

AI in cyber security is significantly more than a negligible utilization of the calculations. It tends to be utilized to break down digital dangers better and react to security episodes. There are a couple of other huge advantages of AI, as follows:

- Detects malignant exercises and stops digital assaults
- Analyses portable endpoints for digital dangers – Google is as of now utilizing AI for the equivalent
- Improves human examination – from pernicious assault location to endpoint insurance
- Uses in computerizing everyday security errands
- No zero-day weaknesses

3.4.7 AI Adopters Moving to Make a Move

PC-based knowledge has quite recently been grasped to strengthen the security system of organizations. There are different veritable models where AI reasoning controlled courses of action are by and large improving cyber security [14].

- Gmail uses AI to square a hundred million spams in a day. It has developed a system to filter through messages and offer a without spam condition gainfully.
- IBM's Watson scholarly getting ready uses AI to recognize advanced perils and other cyber security plans.

- Google is using Profound Learning artificial insight on its Cloud Video Knowledge stage. On this stage, the chronicles set aside on the labourer are explored reliant on its substance and setting. The recreated knowledge figuring's send security alerts at whatever point something questionable is found.
- Belbin stage uses AI consciousness controlled risk desires to guarantee the IT establishment against data and security enters.

3.5   AI Impacts on Cyber Security

There is as of now a major discussion going on about whether computerized reasoning (Artificial intelligence) is a fortunate or unfortunate thing as far as its effect on human life. With an ever-increasing number of endeavours utilizing artificial intelligence for their necessities, it's an ideal opportunity to break down the potential effects of the execution of Artificial intelligence in the digital security field shown in Figure 3.5 [5].

1. Quicker Discovery and Reaction Times
   Computerized reasoning can quicken the acknowledgment of bona fide issues, rapidly cross-referencing different alerts and wellsprings of security data. Human computerized security pros will even now make the methodologies the necessities of the scenes to be dealt with. In any case, it tends to be further helped by simulated intelligence structures that suggest plans for improving responses.

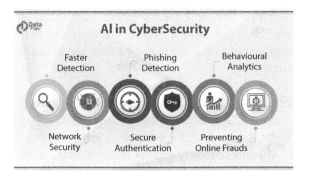

Figure 3.5 Effects of AI in cyber security.

2. System Security

   Two huge bits of framework security are the creation of security procedure and comprehending an organization's framework geology. Ordinarily, both of these activities are exceptionally repetitive. We can use Artificial intelligence to accelerate these techniques, which it does by watching and learning framework traffic structures similarly as suggesting security courses of action. That doesn't simply save time, it also saves a lot of effort and resources which we can rather apply to regions of a mechanical new development and progress.

3. Phishing Discovery and Counteraction Control

   One of the most generally used computerized attack strategies, where software engineers endeavour to pass on their payload using a phishing ambush, is phishing. Phishing messages are normal; one in each 99 messages is a phishing attack. Fortunately, computer-based intelligence ML may accept an essential activity in preventing and diverting phishing ambushes.

   PC-based knowledge ML can perceive and follow more than ten thousand dynamic phishing authority and respond and remediate significantly faster than individuals can. Also, simulated intelligence ML works at separating phishing risks from all over the world. There is no confinement in its cognizance of phishing endeavours to a specific land or area. PC-based knowledge has made it possible to distinguish between a fake site and a genuine one quickly.

4. Secure Validation

   Passwords have reliably been astoundingly fragile with respect to security. Additionally, they are frequently the main barrier between digital hoodlums and our records. The essential way secure confirmation can be cultivated is by physical unmistakable evidence, where artificial intelligence uses different components to recognize a person. For instance, a wireless can use exceptional finger impression scanners and facial affirmation to allow you to sign in. The strategy behind this includes the program looking at essential data that centres on your face and fingers to see if the login is valid. Other than that, simulated intelligence can examine various segments to choose whether a specific customer is endorsed to sign in to a specific gadget. The tech analyses factors like the way you enter keys, your creating speed, and your mix-up rate while composing.

3.6 The Positive Uses of AI Based for Cyber Security

- Biometric logins are logically being used to make secure logins by either looking at fingerprints, returns, or palm prints. This can be used alone or identified with a mystery expression and is starting to be used in most new PDAs. Huge associations have been the overcomers of security enters which compromised email addresses, singular information, and passwords. Computerized security authorities have emphasized various occasions that passwords are frail against cube attacks, bartering singular information, charge card information, and government overseen investment funds numbers. These are for the most part reasons why biometric logins are a positive man-made intelligence commitment to digital security [15].
- Artificial information can additionally be utilized to isolate risks and other possibly malignant exercises. Standard structures can't stay aware of the sheer number of malware that is made each month, so this is a possible locale for PC-based understanding to step in and location this concern. Cyber security organizations are training computer-based intelligence frameworks to distinguish infections and malware by utilizing complex calculations so simulated intelligence would then be able to run design acknowledgment in programming. Computer-based intelligence frameworks can be set up to perceive even the most diminutive acts of ransom ware and malware attacks before they enter the structure and a while later isolate them from that system. They can in like manner use judicious limits that beat the speed of customary strategies.
- Structures that sudden spike in demand for computer-based intelligence open capability for common language preparing which gathers data naturally by going through articles, news, and studies on digital threats. Potential for regular language getting ready which accumulates information normally by experiencing articles, news, and studies on advanced threats. This information can give understanding into irregularities,

computerized ambushes, and expectation techniques. This grants computerized security firms to stay invigorated on the latest perils and timeframes and manufacture responsive strategies to keep organizations secure.

- Computer-based intelligence frameworks can likewise be utilized in circumstances of multifaceted confirmation to give access to their clients. Various clients of an organization have various degrees of verification benefits which likewise rely upon the area from which they're getting to the information. At the point when computer-based intelligence is utilized, the validation system can be much progressively powerful and continuous and it can change get to benefits dependent on the system and area of the client. Multifaceted verification gathers client data to comprehend the conduct of this individual and make an assurance about the client's entrance benefits [10].

To utilize artificial intelligence to its fullest abilities, it's critical that it's completed by the benefit advanced security firms who think about its working. In spite of the fact that previously, malware ambushes could occur without leaving any sign on which weakness it abused, reproduced knowledge can step in to guarantee the advanced security firms and their clients from attacks regardless, when there are various talented ambushes occurring [16].

3.7   Drawbacks and Restrictions of Using Computerized Reasoning For Digital Security

- The advantages sketched out above are just a limited quantity of the capacity of AI consciousness in supporting computerized security; there are in like manner imperatives which are keeping PC-based insight from transforming into a standard instrument used in the field. To manufacture and keep up and computerized reasoning system, associations would require a massive proportion of advantages including memory, data, and handling power. Moreover, in light of the fact that recreated knowledge systems are set up through learning educational files, computerized security firms need to get their hands on a wide scope of enlightening files of malware codes, non-poisonous codes, and idiosyncrasies.

Procuring these exact educational assortments can take a really delayed time frame and resources which a couple of associations can't oversee.

- Disadvantage: Another drawback is that software engineers can in like manner use artificial knowledge themselves to test their malware and progress and overhaul it to possibly become artificial intelligence proof. Honestly, a recreated knowledge affirmation malware can be incredibly perilous as they can pick up from existing artificial insight gadgets and develop additionally created attacks to have the alternative to enter regular computerized security programs or even PC-based insight helped systems.

3.8 Solutions to Artificial Intelligence Confinements

Knowing these imprisonments and drawbacks, obviously PC-based insight is a long way from transforming into the fundamental digital security course of action. The best approach in the meantime is get customary procedures together with reproduced insight instruments, so organizations ought to recall these courses of action when working up their computerized security strategy:

- Employ a cyber-security firm with specialists who have comprehension and capacities in a wide scope of highlights of digital security.
- Have your computerized security team test your structures and frameworks for any potential openings and fix them immediately.
- Use channels for URLs to square dangerous associations that conceivably have a disease or malware.
- Install firewalls and other malware scanners to make sure about your systems and have these constantly revived to organize updated malware.
- Monitor your dynamic traffic and apply leave channels to restrict such a traffic.
- Constantly review the latest computerized risks and security shows to get information about which perils you should manage first and develop your security show in a similar way.

- Perform customary audits of both gear and programming to guarantee your structures are strong and working.

Following these methods can help moderate enormous quantities of the threats related to advanced attacks, but it's basic to understand that your organization is still at risk for an ambush. Thusly, balance isn't adequate and you should in like manner work with your advanced security term to develop a recovery technique.

As the capacity of artificial intelligence is actuality researched to help the computerized security profile of an association, it is also subsistence made by software engineers. After all it is so far being made and its inert limit is far from reach, we can't yet know whether it will one day be valuable or horrible for computerized security. In the interim, it's huge that organizations do as much as could be expected with a mix of traditional strategies and PC-based knowledge to keep consistent over their advanced security strategy.

3.9   Security Threats of Artificial Intelligence

All cyber-attacks can be classified within the most widely recognized terms of accessibility and trustworthiness, interweaved to shape three primary bearings shown in Figure 3.6.

Espionage or Secret activities, which regarding cyber security implies gathering experiences about the framework and using the acquired data for their own benefit or plotting further developed assaults. At the end of the

Figure 3.6 Security threats.

day, a programmer can utilize an ML-based motor to penetrate down and become familiar with the contents like data sets.

Sabotage (Harm) by the resolve to cripple usefulness of a computer-based intelligence framework with inundating simulated agility with solicitations, or model change.

Deception (Misrepresentation) which in artificial intelligence signify misclassifying errands, for instance, presenting off base information in the preparation dataset (data harming) or connecting by a framework at knowledge or conception phase.

3.10   Expanding Cyber Security Threats with Artificial Consciousness

People are looking every day to change the way in which society copes with advancement, and one of the latest developments in the field of software engineering is AI. Various organizations are researching the use of computer-based intelligence and AI to see how to secure their frameworks against digital assaults and malware. However, given their inclination of self-learning, these artificial intelligence systems have now additionally accomplished a level where they can be prepared to be a danger to structures, i.e., go into hard and fast assault mode.

It is for all intents and purposes inescapable that we'll see a rise in usage for artificial intelligence in our everyday life. Be that as it may, much like various progressions in the tech showcase, this also opens up a totally unique avenue of developed misapplication from the digital aggressors. There is the unequivocal likelihood for a degree of abuse beyond what we've seen up until this point, mostly as the tech continues pushing ahead [8].

Let's discuss how AI is affecting the cyber security negatively.

1. Hackers get an edge with simulated intelligence:
   Cyber security masters acknowledge that the introduction of man-made brainpower is significant for con artists and programmers as well. Cybercriminals can use computerization to mark the direction toward discovering new weaknesses that they can exploit quickly and create several problems.

   Scientists and experts are alarmed about the threat simulated intelligence innovation models pose for cyber security, that for the most part keeps our PCs and data — and organizations' and governments' PCs and data — safe from cybercriminals.

The fear for a few is that computer-based intelligence will convey with it the start of new sorts of advanced dangers that evade normal techniques for countering attacks. For example, AI can make assaults on content at a pace and multifaceted level that generally can't be matched by individuals. Here are the further dangers to digital security through computer-based intelligence.

2. Weaponizing automatons and vehicles:
   Specialists from Cambridge and Oxford anticipated that artificial intelligence might be rehearsed to hack self-driving vehicles and automatons, creating the possibility of purposeful vehicle accidents and maverick bombings. For example, Google's Waylon independent vehicles apply deep learning, and that framework could be outmanoeuvred into perceiving a stop sign as a green light, which could cause a fatal accident. Further, as the intricacy of artificial intelligence and cybercriminals increases, so will the chances of continuous assaults. For example, a couple of programmers can incorporate automatons into a multitude that might be fixed with explosives to cause deadly assaults. It is computerized reasoning innovation that engages cybercriminals to program these assaults all the more adequately and to associate the automatons with essential information.

3. Bot Programmers:
   We value speaking with chatbots without recognizing the measure of information we are giving them. Similarly, chatbots can be modified to keep up conversations with clients in a way that will persuade them to reveal their monetary or individual data, associations, and so forth. In 2016, a Facebook bot addressed itself as a companion and tricked 10,000 Facebook clients into a malware establishment. Once the malware was endangered, it held onto the Facebook record of those clients. Simulated intelligence-enabled botnets can weaken HR through telephone and online support. Most of us are using computer-based intelligence conversational bots, for instance, Amazon's Alexa or Google Aide but we don't grasp the extent of information they have about us. Being an IoT-driven innovation, they can for the most part hear even the private conversations happening around

them. Also, some chatbots are ineffectively arranged for secure data transmissions, for instance, Transport Level Verification (TLV) or HTTPS conventions can be adequately used by programmers.

- Spear-phishing is getting simple:
  Artificial consciousness in security assaults will in like manner make it easier for low-level digital assailants to control complex interferences by simply processing effortlessly. Developers routinely win by scaling their errands. The more people they follow phishing plans or, the more frameworks they investigate, it is practically certain they will to get to where they want to go. AI outfits them with a way to deal with scale to much more elevated level, through mechanization of the objectives and conveying mass assaults. A key point of reference where cybercriminals are using computer-based intelligence to dispatch an assault is through lance phishing. Computerized reasoning systems with the help of AI models can save themselves a lot of trouble by emulating individuals by making influencing counterfeit messages. Applying this method, programmers can use them to perform more phishing assaults. Programmers can moreover use computer-based intelligence to make malware for misleading projects or sandboxes that attempt to discover rebel code before it is sent in the arrangement of an association.

  Furthermore, simulated intelligence-based phishing tricks are only the origin. By using AI, digital aggressors could look for potential weaknesses and automatize scope of their potential casualties. The comparable innovation could serve them in sufficiently dissecting the man-made intelligence-based digital guard systems and produce new sorts of malware that could slide through them.

- Malicious defilement
  Associations' simulated intelligence exercises present an assortment of expected weaknesses, fusing malevolent debasement or preparing information, utilization, and portion setup. No industry is safe, and there are various groupings in which artificial intelligence and AI starting at now have a duty and as such present extended dangers. For example, Visa trick may get straightforward. Additionally, security, condition, and wellbeing frameworks might be

risked that control digital physical devices which supervise train directing, traffic stream, or dams [9].

- Social system planning:
  Other artificial intelligence-based dangers will incorporate significant level long-range interpersonal communication planning. For example, computer-based intelligence fueled devices that will look further into long-range interpersonal communication stages will enable psychological militants to identify the correct city and human targets and work even more effectively.

  In this manner, cyber security experts and resistance offices should work united to perceive such perils and make arrangements.

- Home assaults
  Various portions of our own lives are already robotized – with IoT-associated devices and remote helpers. Be that as it may, this requires a great deal of individual data live in the cloud. This complicated arrangement of associations will make new elements of weakness, with digital dangers hitting a great deal closer to home — for example, fear-based oppressors, maverick governments, and programmers could target a web-connected clinical device.

  Quite a while from now, alert frameworks and locks may not be adequate to protect us at home. Maverick bots could check frameworks, chasing down weaknesses. Thusly, an appealing objective is a vulnerable one. It infers anyone could transform into a possible target.

3.11   Artificial Intelligence in Cybersecurity – Current Use-Cases and Capabilities

Artificial knowledge has made a couple of advances in the cyber security territory and a couple of artificial brainpower dealers assurance to have impelled things that use artifical insight to help shield against computerized threats. At Emerj, we've seen various cyber security dealers offering PC-based knowledge and artificial intelligence-based things to help perceive and oversee advanced threats. In the United States, the Pentagon created the Joint Artificial Intelligence Center (JAIC) to help defend U.S. critical infrastructure from malicious cyber activity [11].

In this article, we deal with some of the more typical use-cases for artificial intelligence in cyber security, where there has been some proof of true business use. In particular, we discuss:

3.11.1   AI for System Danger Distinguishing Proof

- AI Email Checking
- AI-based Antivirus Programming
- AI-based Client Conduct Demonstrating
- AI for Battling AI-based Threats

We start our examination of computer-based intelligence in the cyber security space with a clarification for why simulated intelligence is such a solid match for cyber security.

3.11.2   The Common Fit for Artificial Consciousness in Cyber Security

For a business protecting their data, arranging security is essential, and even small businesses may have numerous applications running, all of which need to have different security approaches approved. Human professionals may take a couple of days to weeks to totally appreciate these game plans and guarantee the safety implementation is sound.

Cyber security intrinsically includes tedium and dullness. This is on the grounds that distinguishing proof and evaluation of cyber threats require scouring through enormous volumes of information and searching for strange information focuses. Organizations can utilize the information gathered by their current standards-based system security programming to prepare computer-based intelligence calculations towards recognizing new cyber threats.

Understanding the outcomes of the assault and the reaction required from the organization likewise requires further information examination. Artificial intelligence calculations can be prepared to make certain predefined strides in case of an assault and after some time can realize what the best reaction ought to be through contribution from cyber security topic specialists.

Human security specialists can't match the speed and scale at which computer-based intelligence programming can achieve these information investigation undertakings. Furthermore, artificial intelligence-based cyber security information investigation programming can finish the assignment with reliably higher precision than human experts. Huge scope information examination and irregularity recognition are a portion of the regions where artificial intelligence may include esteem today in cyber security.

Numerous cyber security interruptions normally work over the venture arrange observing the information going all through the system is one approach to identify cyber security dangers. Observing every "bundle" of information that is a piece of the venture systems interchanges is practically inconceivable for human examiners to screen precisely [12].

AI-based programming can conceivably utilize various procedures, for example, measurable examination, catchphrase coordinating, and peculiarity identification to decide whether a given bundle of information is distinctive enough from the benchmark of information parcels utilized in the preparation dataset.

The entirety of this appears to demonstrate that computerized reasoning is currently beginning to be viewed as a compelling instrument to increase genuine preferences against fraudsters and programmers.

3.11.3   Artificial Intelligence for System Danger ID

Undertaking system security is basic for most organizations, and the hardest part about building up great system cyber security forms is seeing all the different components associated with the system geology. For human cyber security specialists, this implies tedious work in following all the correspondences going all through the venture arrange.

Dealing with the security of these undertaking systems includes distinguishing which association demands are real and which are endeavouring uncommon association conduct, for example, sending and accepting enormous volumes of information or having irregular projects pursuing association with a venture arrange.

The test for cyber security specialists lies in recognizing which parts of an application, regardless of whether on the web, versatile stages, or applications that are being developed or testing, may be malevolent. Recognizing the malevolent applications among a great many comparable projects in a huge scope venture organize requires colossal measures of time and human specialists are not generally exact [13].

Artificial reasoning based framework security programming can screen all drawing nearer and dynamic framework traffic in order to perceive any questionable or unusual models in the busy time gridlock data. The data being alluded to here is commonly too voluminous for human digital security authorities to unequivocally orchestrate risk scenes.

3.11.4   Artificial Intelligence Email Observing

Venture firms comprehend the significance of observing email interchanges so as to forestall cyber security hacking endeavours, for example, phishing. AI-based observing programming is presently being utilized to help improve the discovery exactness and the speed of recognizing cyber threats.

A few distinctive artificial intelligence innovations are being utilized for this utilization case. For example, some products use PC vision to "see" messages to check whether there are highlights in the email that may be characteristic of dangers, for example, pictures of a specific size. In different cases, normal language handling is utilized to peruse the content in messages coming in and leaving the association and recognize expressions or examples in text that are related with phishing endeavours. Utilizing inconsistency location programming can help recognize if the email's sender, beneficiary, body, or connections are dangers.

This utilization case again features computer-based intelligence's qualities with enormous scope information examination. It isn't hard for a human worker to peruse an email and distinguish dubious highlights; however, doing as such for many messages sent and received in a huge association on a daily basis is essentially impossible. Computer-based intelligence programming can rather peruse all the approaching and cordial messages and report the most probable instances of cyber security dangers to security staff.

For example, cases to give email checking artificial intelligence programming that can enable budgetary firms to forestall misled messages, forestall information penetrates and phishing assaults. The organization's product likely uses common language preparing and abnormality identification in various strides so as to distinguish which messages are likely to be cyber security dangers [14].

3.11.5   Simulated Intelligence for Battling Artificial Intelligence Dangers

Organizations need to develop the speed with which they identify cyber threats on the grounds that programmers are currently utilizing artificial intelligence to possibly invent resolutions of section in huge professional classifications. In this way, assigning simulated intelligence software development to make preparations for man-made intelligence enlarged hacking endeavours could chance hooked on a fundamental bit of digital safety officer shows later on.

In recent years, organizations around the globe have surrendered to cyber threats and ransom ware assaults, for example, Winery and notpetya. These categories of attacks spread quickly and influence countless PCs. All things considered, the felons responsible for these types of attacks can apply computer-based intelligence revolution later on. The bit of leeway that simulated intelligence could give these programmers is like what artificial intelligence suggestions in establishments: fast versatility.

Cyber Security Merchant Gathering assault claims their retreat software development, Bird of prey Stage, utilizes artificial intelligence to prepare for such ransomware threats. The creation allegedly uses inconsistency discovery for end-point retreat in immense commercial classifications.

3.11.6   The Fate of Computer-Based Intelligence in Cyber Security

Artificial knowledge use in digital security structures can regardless be named as early at this moment. Associations need to ensure that their structures are being set up with commitments from digital security pros which will improve the item at recognizing real advanced ambushes with evidently more precision than regular digital security systems.

Associations need to understand that these systems are simply in a similar class as the data that is being dealt with to them. Simulated intelligence systems are ordinarily extensively elevated to be "garbage in, garbage out" structures, and a data-driven approach to manage PC-based insight adventures is principal for continued with progress.

The one test for associations using essentially artificial consciousness-based digital security disclosure methods is to lessen the amount of fake positive revelations. This may conceivably get less complex to do as the item acknowledges what has been marked as counterfeit positive reports. At the point when a standard of direct has been created, the counts can hail authentically significant deviations as peculiarities and prepared security inspectors that additional assessment is essential.

Digital security submissions are between the maximum notable AI solicitations nowadays. This is in tremendous part due to the way that these applications rely upon peculiarity acknowledgment which AI models are extremely suitable for. In addition, most enormous associations may starting at now have existing digital security gatherings, thing headway spending plans and IT establishment to manage a great deal of data.

3.12   How to Improve Cyber Security for Artificial Intelligence

This strategy brief investigates the key issues in endeavouring to improve cyber security and security for artificial consciousness just as jobs for policymakers in helping address these difficulties. Congress has just demonstrated its enthusiasm for cyber security enactment focusing on particular sorts of innovation, including the Internet of Things and casting a ballot frameworks. As simulated intelligence turns into a progressively significant and generally utilized innovation across numerous areas, policymakers will discover it progressively important to think about the crossing point of cyber security with based of AI. I portray a portion of the issues that emerge around there, including the trade-off of artificial intelligence dynamic frameworks for noxious purposes, the potential for foes to get to private artificial intelligence preparing information or models, and strategy proposition planned for tending to these worries.

Securing Artificial Intelligence Decision-Making Systems:
One of the critical security threats to artificial reasoning systems is the potential for foes to deal with their dynamic techniques so they don't make choices in the manner that their engineers would expect or need. One way to achieve this would be for adversaries to truly accept accountability for a man-made knowledge structure with the objective that they can pick what yields the system produces and what decisions it makes. Then again, an attacker may endeavor to affect those decisions even more straightforwardly and in an indirect manner by passing on toxic data sources or getting ready data to a PC-based knowledge model.

For instance, an enemy who needs to deal a self-administering vehicle with the objective that it will undoubtedly get into a setback may abuse shortcomings in the vehicle's item to choose driving decisions themselves. In any case, indirectly getting to and abusing the item working a vehicle could prove irksome, so taking everything into account an adversary may endeavor to make the vehicle ignore stop signs by demolishing them in the region with splash painting. As such, the PC vision count would not have the alternative to recall them as stop signs. This strategy by which adversaries can make PC-based knowledge structures submit blunders by controlling wellsprings of data is called hostile air. Researchers have found that little changes to cutting-edge pictures that are indistinct to the regular eye can be satisfactory to make recreated insight computations absolutely misclassify those photos.

An elective way to deal with controlling sources of info is information harming, which happens when foes train an artificial intelligence model on off base, mislabelled information. Pictures of stop signs that are marked as being something different so the calculation won't perceive stop signs when it experiences them out and about is a case of this. This model harming would then be able to lead an artificial intelligence calculation to commit errors and misclassifications later on, regardless of whether an enemy doesn't approach legitimately control the sources of info it receives. Even just specifically preparing a man-made intelligence model on a subset of accurately named information might be adequate to bargain a model with the goal that it settles on off base or startling choices.

3.13   Conclusion

Today, people are living in advanced societies where hard and fast data or information is kept up in electronic/online structure. The information may be related to individual life, cash-related trades, authorized advancement or whatever other information which is critical in nature. Surely, bundles of information have been presented on individual correspondence regions without understanding the safety risks. This is the characteristic item for computerized hooligans as the information is in open access. Digital security isn't only an issue related to a person. It is for an organization as well. Each moment one must be able to guarantee data or information on relational connection goals, and the information related to bank trades must have enough security endeavors. Cyber security isn't only a data innovation division or an issue or obligation concerning individuals in a similar office. It is the activity of each worker and even clients of the association. According to Google web crawler, people are going online like clockwork every minute of every day. So how might we safeguard it? General Date Protection Regulation (GDPR), a regulation in European Union law, gives individuals more control over their personal data. However, there are associations which have endured cyber-attacks, are going to endure and may have endured them, yet don't have the foggiest idea about what to do about it. To discover better answers on this we need computer-based intelligence strategies. Understanding the connection between computer-based intelligence or science that can copy individuals and cyber security that is a fundamental requirement for everything is the way to accomplishment in business today. There are systems available to guarantee information on the internet, such as mystery word security, confirmation of data, malware scanners, firewalls, antivirus programming, etc. Through realizing authentic advanced ethics, a larger number of computerized ambushes can be countered. However, computerized infringement or attacks are continuing to proliferate down various paths over time. There is no perfect course of action or across the board answer for computerized bad behaviors/ambushes, yet the latest devices are able to constrain them in order to ensure security in an advanced world.

References

1. B. S. Fisher, S. P. Lab, (2010) *Encyclopedia of Victimology and Crime Prevention*, SAGE Publications, Vol. 1, pp. 251, USA.
2. Singh Rathore, P., Kumar, A., & Gracia-Diaz, V. (2020). A Holistic Methodology for Improved RFID Network Lifetime by Advanced Cluster Head Selection using Dragonfly Algorithm. *International Journal of Interactive Multimedia and Artificial Intelligence*, 6 (Regular Issue), 8. http://doi.org/10.9781/ijimai.2020.05.003
3. Bharat Singh, Ravinder Singh and Pramod Singh Rathore. Article: Randomized Virtual Scanning Technique for Road Network. *International Journal of Computer Applications* 77(16):1-4, September 2013.
4. Naveen Kumar, Prakarti Triwedi, Pramod Singh Rathore, "An Adaptive Approach for image adaptive watermarking using Elliptical curve cryptography (ECC)", *First International Conference on Information Technology and Knowledge Management* pp. 89–92, ISSN 2300-5963 ACSIS, Vol. 14 DOI: 10.15439/2018KM19
5. Cerli and D. Ramamoorthy, "Intrusion Detection System by Combining Fuzzy Logic with Genetic Algorithm", *Global Journal of Pure and Applied Mathematics (GJPAM)*, vol. 11, no. 1, 2015.
6. R. Hill, "Dealing with cyber security threats: International cooperation, ITU, and WCIT", *2015 7th International Conference on Cyber Conflict: Architectures in Cyberspace*, pp. 119-134, 2015.
7. Pramod Singh Rathore, "An adaptive method for Edge Preserving Denoising, International Conference on Communication and Electronics Systems, Institute of Electrical and Electronics Engineers, PPG Institute of Technology. (2017). *Proceedings of the 2nd International Conference on Communication and Electronics Systems (ICCES 2017)*: 19-20, October 2017.
8. S. Adebukola, Onashoga, Akinwale O. Bamidele and A. Taofik, "A Simulated Multiagent-Based Architecture for Intrusion Detection System", *(IJARAI) International Journal of Advanced Research in Artificial Intelligence*, vol. 2, no. 4, 2013.

9. S. Dilek, H. Çakır and M. Aydın, "Applications of Artificial Intelligence Techniques to Combating Cyber Crimes: A Review", *International Journal of Artificial Intelligence & Applications (IJAIA)*, vol. 6, no. 1, 2015.

10. P. S. Rathore, A. Chaudhary and B. Singh, "Route planning via facilities in time dependent network," *2013 IEEE Conference on Information & Communication Technologies*, Thuckalay, Tamil Nadu, India, 2013, pp. 652-655. doi: 10.1109/CICT.2013.6558175

11. N. Bhargava, S. Dayma, A. Kumar and P. Singh, "An approach for classification using simple CART algorithm in WEKA," *2017 11th International Conference on Intelligent Systems and Control (ISCO)*, Coimbatore, 2017, pp. 212-216. doi: 10.1109/ISCO. 2017. 7855983

12. S. Singh and S. Silakari, "A Survey of Cyber Attack Detection Systems", *IJCSNS International Journal of Computer Science and Network Security*, vol. 9, no. 5, 2009.

13. Prof. Neeraj Bhargava, Pramod Singh, Abhishek Kumar, Taruna Sharma, Priya Meena, "An Adaptive Approach for Eigenfaces-based Facial Recognition", *International Journal on Future Revolution in Computer Science & Communication Engineering (IJFRSCE)*, December 17, Volume 3 Issue 12, pp. 213–216.

14. S. W. Brenner, (2010) Cybercrime: Criminal Threats from Cyberspace, Greenwood Publishing Group, USA.

4

Botnet Detection Using Artificial Intelligence

Astha Parihar[1]* and Prof. Neeraj Bhargava[2]

[1]M.Tech Scholar, MDS University, Ajmer, India
[2]School of Engineering & System Sciences, Professor MDS University, Ajmer, India

Abstract

Over the last ten to fifteen years botnets have caught the attention of researchers worldwide. One of the dangerous additions to the gallery of malicious software programs is bot malware, popularly referred to as botnets. A botnet is a network of infected host/machines which can be functioning software program robots and are managed with the aid of a human, via one or more controllers. This chapter carried out a comparative evaluation of literature on previous researches and studies on botnet identity particularly using system language techniques. The examination revealed that the gain and popularity of machine learning algorithms in botnet detection stems from the fact that other styles of botnet detection strategies like the intrusion detection gadget had been visible to be externally incompetent. This chapter further proposes a botnet identification version using optics algorithm that hopes to effectively discover botnets and perceive the type of botnet detected by way of addition of latest feature; incorporation of changed traces to pinpoint supply IP of bot master, identification of existence of the kind of services the botnets get right of entry to are areas the proposed solution will cater to.

Keywords: Botnet, botnet-detection, machine learning, supervised learning, unsupervised learning, botnet architecture, extensive botnet detection system (EBDS)

*Corresponding author: asthaparihar9@gmail.com

Neeraj Bhargava, Ritu Bhargava, Pramod Singh Rathore, and Rashmi Agrawal (eds.) *Artificial Intelligence and Data Mining Approaches in Security Frameworks*, (65–86) © 2021 Scrivener Publishing LLC

4.1   Introduction to Botnet

The term botnet is from the expressions automated and network. A bot, once in a while known as a zombie, is a character gadget connected to a web convention (IP) arrangement, typically the web. Generally, this implied registering gadget PCs, PCs, printers, household switch, etc. Had been obligated to turning into a bot. These days in any case, as the Internet of Things (IoT) advances, our family gadgets are in expanding numbers habitually associated with the web. Which implies the up-and-comer rundown of capacity botnet gadgets has essentially been duplicated. Secured now are net cams. After a gadget gets kindled with botnet malware, it might be utilized through its locale network to direct a large number of unapproved and malignant games (B. S. Fisher *et al.*, 2010).

Once a device is installed with a "bot software" via malware infection, "bot herder" can make the bot do anything by issuing commands via a command and control (C&C or C2) server. A botnet typically consists of hundreds or even millions of devices, including PC, Mac, Linux servers, home router, smartphone, etc. The combined resources of these controlled devices can be used to launch destructive or sophisticated attacks like sending billions of spam email, huge bandwidth DDoS and targeted financial fraud.

Botnets have traditionally used HTTP and IBN protocols to communicate with infected botnet clients. To block this communication, network security services can control access to these services and ports. For

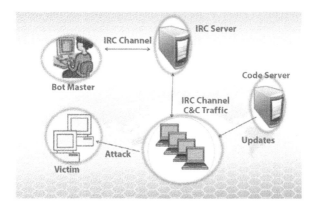

Figure 4.1 Introduction to Botnet.

example, the Firebox can use the WebBlocker Command and Control and Botnet Activity categories to block communication from infected botnet clients on your network to botnet sites over HTTP (Singh Rathore, P. *et al.*, 2020).

Above Figure 4.1 denotes the Botnet overview where all necessary components had been introduced. Botnet herders are entertainers who oversee bots distantly. They arrange and establish command and control (C&C) workers, which fill in as the interface to the bots. Botnet herders are malwares that oversee bots distantly. They associate and establish command and control (C&C) employees, which fill in the interface to the bots. For instance, IBN channels are frequently employed for that reason. After communications are setup, the negotiated hosts are often further prepared and issued updated instructions (Naveen Kumar *et al.*, 2018). They've now emerged as a prepared group of hosts beneath centralized management.

4.2   Botnet Detection

As botnets emerge as greater threats, researchers and safety experts develop special techniques and strategies to resolve the issues. The detection method defines how the answer operates, along with detection by

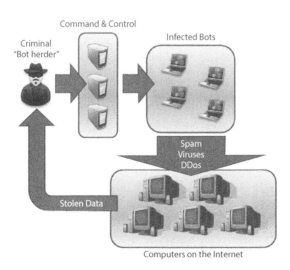

Figure 4.2 Factor of Botnet.

behaviour or signature shown in Figure 4.2 (Cerli and D. Ramamoorthy *et al.*, 2015). Exceptional strategies are primarily based on extraordinary strategies.

Various techniques have been designed for Botnet detection from time to time. Three major techniques of botnet detection are described below:

a) Host-Centred Detection
b) Honey Nets-Based Detection
c) Network-Based Detection

4.2.1   Host-Centred Detection (HCD)

In HCD-based identification methods the examination of machine reactions is done based on specific terms. The overall conduct of the machine is watched and attempt is made to find any sort of variation from the norm (J. Nogueira, 2006). These include framework taking too long to even think about responding to even little activities, taking too long to even consider resolving the call successions, any dubious section in the library, unusual changes in the record frameworks, antivirus not reacting or killing all alone, changes in arrange associations and so on may indicate nearness of a bot. Host-centered discovery strategies are not considered extremely effective techniques on the grounds that such strategies are fit for just one machine and may change from machine to machine shown in Figure 4.3 (H. Dijle *et al.*, 2011).

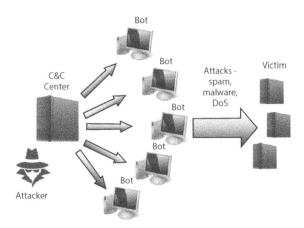

Figure 4.3 Host-centered detection system.

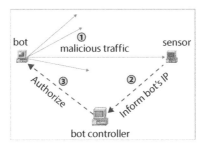

Figure 4.4 Honey nets-based botnet detection.

4.2.2   Honey Nets-Based Detection (HNBD)

Honey nets (sometimes also referred to as honeypots) are for the most part used to contemplate and comprehend botnet highlights and methods; however, they are not generally valuable in identifying bot contamination. Nectar nets are ordinarily utilized to find the goals of bot experts or aggressors. This method is valuable for distinguishing the known bots. The obscure bots and even known bots with slight change in the bot parallels are not identified by this strategy shown in Figure 4.4 (H. Dijle *et al.*, 2011).

4.2.3   Network-Based Detection (NBD)

The system put together location method is based with respect to checking and breaking down the inactive system traffic. This methodology is very useful in recognizing the nearness of botnets in the systems. In this methodology the system information is consistently checked, organize-based correspondences are watched. Any anomalous follow may demonstrate the nearness of some vindictive action. Presently the boasters are keen and apply a large number of code confusion methods. Despite the fact that the malevolent code is jumbled and circumvented by malware discovery programming, the parcels are as yet present in the system, and can be additionally followed by applying different methods.

4.3   Botnet Architecture

This part of the chapter shows some key features of botnets. A botnet is a community including machines under the straight control of a precise operative known as a botmaster. A botnet can be thought of as a merging of numerous dangers into one. A botnet generally contains botnet servers that build up hundreds and thousands botnets with the network. These hundreds and thousands of botnets are known as small botnets, and botnets which contain millions of clients are known as big botnets.

The term botnet is borrowed from "Robotic Community". It reflects the reality that the bot-clients will behave as robots, and servers, that is, the bot-masters, are used to ship the instructions and fulfill the intention of launching the assaults to a one-center location. Nowadays a single botmaster tackles lots of bot servers by setting up various segment (Ionita and L. Ionita *et al.*, 2013).

The botmaster usually placed conversation with bots the usage of IBN [Internet Broadcast Negotiation] on a faraway command and control server. It consists of five major levels featured inside the conversation from linking of a brand new bot to launching an attack, as described below:

(i)   A brand new available device is attacked and conciliated through copying into it a malicious piece of code. When malicious code is executed the machines searches for C&C slaves, which itself connected to the servers and be a part of 72 botnets. Rallying procedure informs the presence of bot-nets and then botmaster is ready to receive the commands.

(ii) After that it receives commands from the botmaster to carry out a few vicious missions.

(iii) The commands received by way of the machine from the botmaster are then achieved by using the bot consumer.

(iv) The attack is introduced according to the given command.

(v)   The bot patron responds to the botmaster to inform him approximately of the fulfilment of the attack.

4.3.1   Federal Model

In the federal model there's one most significant worker or critical component that is liable for setting up verbal trade a couple of the annoyed and bot customers. The utilisation of this channel the trading of messages and directions is taken area. The pivotal worker is known as a command and control (C&C) worker. Numerous accessible botnets along the edge of sabot, ago bot, robot, etc., use C&C for the explanation of discussion. The basic PC or worker is frequently an amazing PC gadget since it needs to address the entire botnet whose size likewise can fluctuate from certain parcels to numerous heaps of hundreds. It ought to have a high data transfer capacity considering the way that at any single issue it might need to serve numerous bots. In spite of the way that the basic worker is an incredible one anyway it is viewed as an inclined factor of this rendition. There are conventions which may be regularly utilized by C&C to do correspondence and that are HTTP (hyper literary substance move convention) and IBN (J. Raiyn *et al.*, 2014).

4.3.1.1   IBN-Based Protocol

IBN is a convention for real-time web text informing or coordinated conferencing dependent on TCP that can likewise utilize loosened up attachment layer. IBN gives various beneficial abilities. It allows shifting files among customers and the packages on foot on structures. IBN primarily based botnets use centralized command and manipulates form wherein inflamed machines try and set up connection to the IBN server and be a part of the identical channel. In these botnets the C&C server works on IBN company. The IBN protocol is primarily based on customer-server model. It gives flexibility in conversation and is pretty easy to set up. It is one of the most popular protocols to putting verbal exchange amongst botnets (O. Oriola *et al.*, 2012).

4.3.1.2   HTTP-Based Botnets

IBN can be thought of as a starting convention for botnets. IBN won great prominence and a large portion of the botnets worked on IBN, such that a ton of analysts began spending significant time in IBN-based absolutely report. IBN moreover had some demerits. As IBN consists of facts about the port extensive range in advance than taking off an assault, the assault may be detected without difficulty. So the hackers switched to HTTP protocol. This protocol is generally utilized in any category of community. It offers numerous advantages. It has the functionality to hide malicious botnet web page site visitors in regular net website online site visitors which could not additionally be detected through firewall. The completely HTTP-based botnets are easy to shape and execute. There are some botnets which use HTTP protocol and they're restock, clickbot, Zeus, there are styles of HTTP primarily based absolutely genuinely botnets:

1. Resonance centred totally clearly HTTP botnets
2. Knowledge primarily centred definitely honestly HTTP botnets

4.3.2   Devolved Model

A model is said to be decentralized when there's no significant command and manipulation. The custom used by centralized botnets is IBN and HTTP even though these botnets work with several types of protocols. Peer to peer (P2P) is an instance of decentralized version. P2P community of settle machines is sufficient strong to become aware of and ruin. This machine typically uses document sharing networks. In decentralized fashions the botherd has freedom to pick out any bot to distribute commands inside the botnet. All bots can act as customers further to servers. This shape of botnet cannot be taken down through clearly attacking at one thing due to the fact there is no important server to manipulate whole botnet. If one bot is attacked and taken down then one-of-a-kind bots of the botnet will maintain on operating. P2P botnet is extra dynamic and strong than the centralized one. Every bot keeps a few collaboration to the opposite bots of the botnet. The P2P botnet is quite tough to be monitored, taken down and hacked (R. Hill, 2015).

4.3.3   Cross Model

Hybrid architecture is a mixture of both centralized and decentralized structure. States that with hybrid architecture there are two sorts of bots. The client bot and the servant bot. Tracking and detection of botnets having hybrid structure is more difficult than with centralized and decentralized architecture shown in Figure 4.5 (Pramod Singh Rathore *et al.*, 2017).

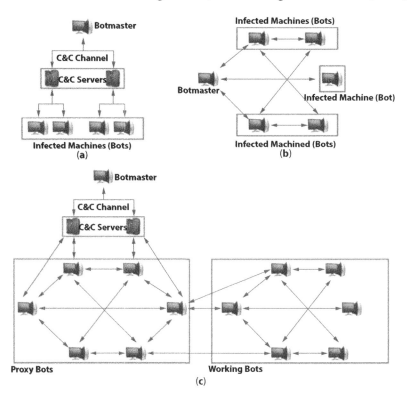

Figure 4.5 Botnet architecture (a) federal model (b) devolved model (c) cross architecture.

4.4   Detection of Botnet

In detection technique solution will be operates along with its nature. Machine learning-based detection strategies are capable of using both technique. One-of-a-kind strategies applied in bot detection include anomaly and area name system.

4.4.1   Perspective of Botnet Detection

Mark-based strategy looks for distinct information on a bot or then again attributes identified with bot, such as Traffic, might also appear like. Signature based totally approach searches for best information of bot or related to bot traits, such as Site visitors, can also look like. This methodology is utilized to specific developments including a specific rules or administration. This kind of methodology turns to specific and particular, something outdoor the desired extension will move uncharted. This technique can be too powerful in opposition to identified botnets, but now this is not useful for anonymous bots and is extra prone to elusion strategies.

Detection-based approach is based mostly on bot behavior and consists of describing a version for a manner in which botnets usually function. The simplicity of this approach makes it feasible to seize new or inconspicuous bots, but, very elegant and the faux outstanding charge might turn out to be immoderate. In behavioral techniques researchers make assumptions based totally on observations about center behaviors of botnets. For botnet detection, the number one assumption sooner or later of techniques is that bots carry out in a cooperative manner, engaging in some form of group interest at various levels of the botnet existence cycle. In which particular statistics of a particular bot drives signature based truly approaches a smooth definition of bots' behavior is at the center of behavioral strategies.

4.4.2   Detection (Disclosure) Technique

ABD (Anomaly-Based Detection) strategies aim to come upon bots dependent on amazing web sports, which encompass abnormally excessive transportation, large potential and uncommon port sports action. ABD techniques takes a conduct strategy to bot identification, hence, it can choose up astonishing games or conduct for obscure bots.

Technique based on DNF performs facts formed with the resource of botnets. C&C conversation channels are unique to bot malware; bots have interaction with C&C servers via those channels. To get entry to those hubs, bots execute DNF queries. The objective of DNF-based strategies is to get uncommon DNF site guests to see bots.

AI basically subject to introduction frameworks had been viewed as the best at distinguishing botnets.

Its efficiency lies in its potential to come to be aware of bot associated site visitor's inner regular traffic. That is an assignment for different strategies as bots utilize ordinary conventions to makes C&C report. But, machine learning detection calls for an appropriate quantity of schooling cases and properly described capabilities to be operative. As the point of interest of this chapter, device learning will be mentioned in greater detail in the following section.

4.4.3   Region of Tracing

Distinguishing bots dependent on association venture movement accept composed games by means of bots inside the equivalent botnet. Through comparable website traffic styles, the aim is to understand all of the bots within the community based mostly on their collective movement as a substitute that their character operations. In comparison to institution based totally detection, man or woman swarms are labeled based on their character activities and traits spite of the interest of the institution they are probably part of.

4.5   Machine Learning

Machine learning is an application of AI that aims to increase system that is able to analyse from the beyond experience. In device learning, past facts are given as an input to device gaining knowledge of algorithm to accumulate styles that could exist with a purpose to build a model that shows the records. On the focal point of ML are measurable and computational thoughts got from rules that exist in loads of orders which comprise manmade consciousness, theory, measurements thought, science, psychological innovative skill, computational multifaceted nature and control statute.

The goal of ML is to form a version based at the statistics given. This version describes the styles that exist in the records which ought as a manner to make knowledgeable picks given new (unseen) facts.

4.5.1   Machine Learning Characteristics

Machine learning technique is used for tracing the botnets. The features decided the shape of version that is fashioned. Functions are capable to show the behaviours or goal of specific traits. The ML-method chosen will affect how the model behaves; one approach can also additionally further-more create a version whose maximum essential challenge is how partic-ular bots interact with every other while every other model troubles itself with how individual bots characteristic.

To extract the optimal division of variables through all feasible vari-ables that denotes maximum precise information, a method called capa-bilities choice is used. In botnet location, the motivation behind the activity decision method is to select a subset of capacities while in tran-sit to top mark depict the direct of bots or the best bot being bounced. Competencies decided on will be based totally on the form of infor-mation getting used. Various type of query lookup may be a function from DNF statistics, deliver and vacation spot ip for internet go with the flow records, checksum are functions from packet top statistics. For ML based totally detection maximum researchers decided on net flows (N. Bhargava *et al.*, 2017).

Examples of waft-diploma capabilities are: go with the flow time, com-mon byte in keeping with packet in keeping with drift, who indicated the relationship between consumer and server. The capabilities decided on will useful resource a specific approach. Waft degree capabilities will assist a behavioural technique, packet degree competencies that seize specific trends will assist a signature-based technique. The basic hypothesis for AI fundamentally based botnet discovery is that, bots create precise styles invisible in web site visitors or purchase the appliance sports activities. For this reason, implementing some form of ML technique, one may be able to find those patterns to efficaciously stumble on malicious activity.

4.6   A Machine Learning Approach of Botnet Detection

Originally, maximum strategies installed to thwart botnets were reactive, decreasing their effectiveness appreciably. Great researches have been made recently into more proactive strategies aimed toward tackling bot-nets. The proactive methods delved into the dynamics of botnets in a bid to understand their cycle, functions, structure, design, and attack pattern as well as automatic and actual time methods to identification and detection of botnets.

The fundamental hypothesis for system studying primarily dependence on botnet detection is that bots create particular designs hidden in a com-munity movement or patron device events. Enforcing device studying algorithms may want to assist in finding those hidden patterns to efficaciously hit upon malicious activity.

Numerous processes have been using an array of MLAs deployed in diverse setups. These methods recruit several principles of traffic analysis focused on various traits of botnet network activity. Moreover, present-day detection strategies were evaluated using special assessment methodologies and facts units. The exceptional variety of diverse detection solutions introduces the want of a comprehensive approach to summarizing and evaluating present clinical efforts, with a goal of information the challenges of this elegance of detection techniques and pinpointing possibilities for the future.

Some authors have attempted to summarize the sphere of botnet safety via a series of survey papers. In parallel, several authors have summarized clinical efforts on botnet detection by means of offering novel taxonomies of detection methods and a number of the maximum prominent strategies. The authors have recounted the capacity of gadget mastering based totally tactics in supplying efficient and powerful detection. Machine Learning (ML) techniques used in botnet detection can be discussed underneath the two main classifications of machine leaning, particularly supervised and unsupervised device learning strategies.

4.7   Methods of Machine Learning Used in Botnet Exposure

This phase evaluated the contribution of every competencies in virus detection based on how they have been used. The assessment of each technique will be separated as follows: first, a definition of the approach and second, a brief description of how the method has been used.

4.7.1   Supervised (Administrated) Learning

In this type of learning, fashions are constructed at labelled learning records. The aim is to make a version (task h) that represents the records, defined by a task (h) that charts enter variables x to their suitable goal y, this characteristic is every so often denoted as the speculation h(x). There is a comparison between supervised learning strategies based on labelled information. Supervised device learning understanding of troubles might be classified as reversion or type. For reversion issues, the labelled purpose prices constitute quite a range of values. For this problem, input variables are assigned to schooling primarily based on styles represented in the records. Class procedures are involved by the connection among group tag and input variables. Tracing of botnet is an instance of such an issue, wherein we're attempting to choose what class package of bundle might be doled out to, i.e., botnet or not botnet site guests.

4.7.1.1 Appearance of Supervised Learning

In 2006, a network-based bot-net detection-based approach was presented. The writers ensured an assessment of three distinct gadget studying techniques for figuring out IRC bot-nets. Detection was achieved in stages; the primary phase classifies site visitors primarily based on IRC traffic. The second one phase classifies IRC chat flows.

IRC bots is another approach of botnet detection. This method is divided into five levels. In the first level, flows which can be maximum probably to not require C&C records are sifted through based totally on understanding of IRC bots, behavioural styles and traits in flow. The second phase makes use of supervised learning to become aware of suspicious traffic flows. The 0.33 degree associations coast dependent on comparable predefined attributes. The gatherings are then surpassed to the fourth stage that utilizes topological evaluation to choose streams with the indistinguishable regulator. The streams with the indistinguishable regulator are then analysed to check whether they might be an aspect of a botnet or not (P. S. Rathore et al., 2013).

The third technique is introduced for figuring out P2P bots. The primary level of this method brings out the feature extraction. On this diploma, unique functions that can be used to symbolise P2P bots are take out from the visitor's runs. The competencies of these flows are handed to the second one diploma in which supervised studying set of regulations is used to classify every go together with the drift. A new approach that is botnet detection system which is used to classify both activities based on bots and another is network client computer.

It includes fine section named as M1 to M5, related with bot activities at the network and separate client. Now the first section, M1, is the Mortal-Growth-Network (MGN) correlation study module. This module detects malicious manner through tracking human method on the host referring to the keyboard and mouse and correlating them with network pastime. The system tests the time distinction among a process generating a mouse click on or keyboard occasion, the deliver of the event also checked whether or not the method is going for walks in the foreground

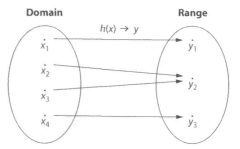

Figure 4.6 Mapping among ML area x and objects y.

at that time is also considered. A small time distinction can also imply that the procedure has been generated with the aid of a human; otherwise this method may be marked as suspicious and forwarded to the M2, M3 and M4. M2 and M3 uses supervised mastering to categorise queried domains names as malicious or benign and classify malicious behaviour on host laptop structures respectively. M4 monitors traffic generated via the suspicious manner at the host's community interface. Incoming packets and alternate price among technique and far-flung website are in comparison. If the alternate ratio is smaller than a predefined cost, bot behaviour is suspected. Ultimately, after each module makes its selection, the correlation engine - M5, combines the results to make the very last selection the usage of a weighted voting scheme shown in Figure 4.6 (S. Gordon, R. Ford, 2006).

4.7.2   Unsupervised Learning

Unsupervised learning in another type of machine learning, belonging with evaluating structures used to discover ways to constitute models in an information set based entirely on enter variables. The goal of such learning is to set a function that detailed hidden pattern in unlabelled facts. The principle target of one in every one of this learner is to set up a feature to provide an explanation for hidden styles in unlabelled facts. The absence of goal values (y), or external environmental evaluation, is what distinguishes unsupervised studying from supervised and reinforcement learning.

The furthermost general type of unsupervised learning is referred to as grouping. This is an unmonitored mastering technique used to find similarity in unlabelled records through grouping them in sections known as clusters.

Seeing that all information factors look the same (unlabelled), the goal of a clustering set of rules is to recognize the relationship among every facts factor and group them accordingly. In exactly the same way, because it pertains to botnet detection, grouping processes must be used to cluster website online traffic of similar traits with a purpose to single out and discover site visitors with malicious cause.

The botminer detection gadget clusters comparable communique site visitors and comparable malicious web page site visitors and executes go bunch relationship to find out the clouds that share same communication designs and comparable cruel interest styles.

4.7.2.1   Role of Unsupervised Learning

K-Means cluster algorithm is proposed for online detection. The approach makes use of network float skills in predefined time windows. The purpose is to institution site traffic primarily based on similarity. The cluster with similarity greater than a predefined threshold might be classed as suspicious; consequently the host associated with the ones go with the flow can be flagged.

Another technique was domain name framework (DNF) carrier. C&C servers are used by bots to DNF look ups. The thought with the guide of the utilization is that, bots separated of the equivalent botnet will utilize DNF contributions also. This strategy utilizes the x-way bunching calculation to association spaces that might be related with a botnet.

Third, a device the ones determine P2P botnets no matter the botnet being presently engaged in malicious interest. The determination of this approach is to find P2P bots by way of identifying C&C communique styles that characterize P2P bots. The device firstly identifies the P2P hosts then P2P bots identifies among those hosts. This method uses drift diploma skills, the machine presumes that P2P nodes create many failed outgoing flows. For each cluster of flows their spot ip is checked and for every ip their bgp prefix is checked. If the range of distinct bgp prefixes are smaller than a predefined quantity they are ignored. To distinguish legitimate P2P website online visitors from bot P2P connections, the authors assume that bots of the identical botnet make use of comparable P2P protocol and network. Additionally they assume that pairs be a part of with the resource of bots which have longer overlaps than that of valid P2P traffic

The usage of the x-manner clustering set of rules, the group glide along same communiqué styles. It contains some additives along three tiers. The primary degree has the A and C-simple video display units that video display units outgoing and inner traffic stream serially. Another phase is prepared from the A and C- surface clustering that clusters site visitors, refined with the aid of manner in their respective video show devices of the preceding degree. The outcomes from those clusters are then exceeded to the 0.33 levels, the skip simple correlate that forms the last choice about hosts that might be an aspect of a botnet. By way of merging the outcomes from the A and C easy groups.

Another method is "Cluster Flow" (CF) which is based on similarity charge. This approach is broken up into three parts, the primary stage examinations characteristic, the second, clusters flows and the 0.33, botnet choice. Inside the early level, capabilities that are derived from the drift charge inside the time durations as a 256 D vector. In the 2nd degree, flows are clustered using okay-approach and x-approach clustering algorithm. The ones clusters are passed to the 0.33 phase wherein the cluster with the lowest widely known deviation is marked as botnet (Neeraj Bhargava *et al.*, 2017.

4.8   Problems with Existing Botnet Detection Systems

Numerous attempts have been made by researchers toward developing frameworks for detection of botnets. Issues have been identified in current botnet detection structures and they phase underneath highlights those troubles and viable solution to address the shortcomings.

A detection system is based on abnormal behaviour of network site visitors. The gadget can hit upon encrypted site visitors and does now not require application layer facts. But, it cannot detect irc bots' verbal exchange on non-preferred ports. Proposed design of detection system that handles actual-time waft could clear up the problem with this system.

Existing peer-to-peer botnet detection system traffic at the gateway is monitoring by using data mining technology. The system's margin consists of that; it really works simplest within a nearby area community surroundings, and will must be allotted to the ISP (Internet Service Provider) degree to be hit upon P2P botnet in a huge scale network. Secondly the lifestyles of NAT era make it tough for the gadget to stumble on P2P flows. The researchers proposed a large-scale community designed for better and greater strong botnet detection.

Spam Preventing System (SPS) for botnet detection is based on support vector machine (SVM). The system handled unsolicited mail blocking by means of separating give up-user machines with valid server machines.

The weak spot of the gadget was that it used a completely small dataset, and turned into determined to be undesirable for small facts set and was observed to be unwanted for small commercial enterprise email servers. The future direction of the research proposed increase in the length and variety of the statistics units.

Proposed, botnet detection machine based totally on machines gaining knowledge of using domain call carrier question facts. Evaluation of the effectiveness of the approach by the use of numerous system gaining knowledge of set of rules and experimental outcomes showed that random forest algorithm produced the satisfactory basic detection accuracy of over 90%. The system no longer handled the effect of domain name function on detection, proposed continuous trying out of the proposed model with larger facts set to assist examine the effects of the DNF feature on enhance detection accuracy.

Due to these problems of existing botnet detection system, Extensive Botnet Detection System (EBDS) has been introduced.

4.9 Extensive Botnet Detection System (EBDS)

The proposed everyday version is an extension of the botnet detection machine constructed. In it, five distinct class techniques, particularly good logic regression, random subspace, randomizable component classifiers, multiclass classifier and random committee, have been applied and evaluated and then the logistic regression classifier changed into recognized as having been most suitable for botnet detection. However, their model became faulted due to the fact that ordinary network usage evolves exponentially in the cyber panorama and after a sure point inside the time the framework would possibly no longer be to distinguish malicious and benign statistics; consequently giving upward thrust to excessive fake fine price. But they proposed that making use of a neural network with deep mastering optimization should produce higher answer by means of accounting for evolving botnets and reducing false positive rates.

This examination consequently proposed a nonexclusive botnet discovery engineering that will convey optics calculation, a neural system calculation with profound learning enhancement to deal with botnet identification in an exponentially developing, digital scene. Assessment of the execution level of the proposed and broadened framework (utilizing optics Algorithm) will be set out upon. The proposed framework will handle notwithstanding botnet recognition; recognizable proof of the sort of botnet and what benefits the botnet is presented to. The proposed botnet recognition framework will likewise catch reaction arrange traffic water-checking. This watermarking will be utilized to follow back to the botmaster in order to pinpoint the IP address of the botmaster. The correspondence among bots and botmaster are bidirectional and intuitive; this is on the grounds that at whatever point a botmaster conveys a message the bots must answer, and the answer must come back to the bot ace. The proposed model targets watermarking the reaction traffic from a bot with the goal that we can in the end follow back to the botmaster shown in Figure 4.7.

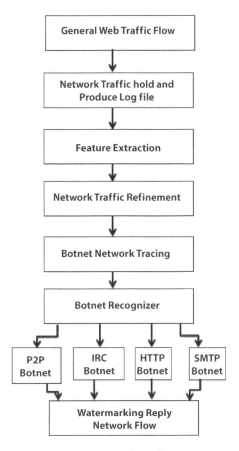

Figure 4.7 Extensive Botnet Detection System (EBDS).

4.10   Conclusion

As bots turned out to be extra undermining, endeavours in the zone increased, creating different techniques of identifying and protecting against botnets. Thus far, machine learning based totally detection approaches have tested to be relatively powerful, yet still have their restrictions. Well-timed recognition, well-timed detection, real-time tracking and flexibility to new threats are issues that remain to be solved. Special ML methods have exclusive powers and flaws as visible within the position they show in bot detection. The measurable premise procedures (for example, the speculation portrayal), stresses itself with the association between the highlights (x) and target (y). While in transit to effectively establish the direct of bots utilizing, this should be depicted with the guide of the highlights chose and therefore expect a couple of explicit information roughly what this lead appears as.

Primarily depending on the properties of SL, this subject has utilized the accuracy of SL strategies to appropriately become aware of bots based totally on a few recognised and precise properties. The accuracy of SL may be pretty powerful in opposition to bot visitors that are trying to camouflage themselves among valid site visitors, given a few particular traits of the malicious visitors. In our survey of the SL strategies we have discovered a commonplace fashion. Other than particular insights about bot traffic found out inside the characteristic area, SL strategies perform flatly. SL strategies may additionally triumph over the covered nature of bots. As seen in Table 4.1, supervised studying strategies are hired for instances wherein some particular feature is understood.

This chapter contains a new botnet detection approach named as Extensive Botnet Detection System (EBDS). This class of detection processes promises automated detection; this is capable of generalize information approximately malicious community site visitors from the to be had observations, for that reason fending off losses of signature-based totally detection procedures which are best able to stumble on known visitors anomalies. To combat botnets in a dynamic terrain, future works on botnet detection, addition of latest capabilities to classify results of a particular kind of botnet. Moreover, they proposed incorporation of modified lines to pinpoint supply IP of botmaster. Identity of life of botnets, what kind of offerings the botnets get entry to is also a proposal for future work. Automation of the entire botnet detection machine might produce knowledge of botnet attributes and might make it simpler to wipe out botnets

Table 4.1 Aspect of machine learning method in botnet detection system.

Method	Region of tracing	Detection outlook	Accurate positive rate	Inaccurate positive rate
Supervised	Particular	Signature (P2P)	98%	2.3%
Supervised	Particular	Signature (IRC)	Not a Number	10 % - 20%
Supervised	Particular	Signature (C&C)	87%	20%
Supervised	Particular	Signature	91%	0.56%
Supervised	Particular	Signature (IRC)	Not a Number	30%
Unsupervised	Cluster	Nature	99%	1%
Unsupervised	Cluster	Nature	100%	20%
Unsupervised	Cluster	Nature	95%	4%
Unsupervised	Cluster	Signature	100%	0.2%
Unsupervised	Cluster	Signature	95%	Not a Number

from a gadget. This proposed device with a bit of luck can have a lesser incidence of fake negative outcomes; similarly the gadget is designed with the goal of most effective detecting a botnet in a community, and it will also pick out what type of botnet it is and similarly classify what offerings and applications the bot has gained admission to.

References

B. S. Fisher, S. P. Lab, (2010) *Encyclopedia of Victimology and Crime Prevention,* SAGE Publications, Vol. 1, pp. 251, USA.

Singh Rathore, P., Kumar, A., & Gracia-Diaz, V. (2020). A Holistic Methodology for Improved RFID Network Lifetime by Advanced Cluster Head Selection using Dragonfly Algorithm. *International Journal of Interactive Multimedia And Artificial Intelligence,* 6 (Regular Issue), 8. http://doi.org/10.9781/ijimai.2020.05.003

Naveen Kumar, Prakarti Triwedi, Pramod Singh Rathore, "An Adaptive Approach for image adaptive watermarking using Elliptical curve cryptography (ECC)", *First International Conference on Information Technology and Knowledge Management* pp. 89–92, ISSN 2300-5963 ACSIS, Vol. 14 DOI: 10.15439/2018KM19

Cerli and D. Ramamoorthy, "Intrusion Detection System by Combining Fuzzy Logic with Genetic Algorithm", *Global Journal of Pure and Applied Mathematics (GJPAM)*, vol. 11, no. 1, 2015.

H. Dijle, N. Doğan, (2011) Türkiye'de Biliδim Suçlarına Eğitimli nsanların Bakıδıll, Biliδim Teknolojiler Dergisi, Vol. 4, No. 2.

Ionita and L. Ionita, "An agent-based approach for building an intrusion detection system", *RoEduNet International Conference 12th Edition: Networking in Education and Research*, pp. 1-6, 26-28, 2013.

J. Nogueira, "Mobile Intelligent Agents to Fight Cyber Intrusions", *International Journal of Forensic Computer Science*, vol. 1, pp. 28-32, 2006.

J. Raiyn, "A survey of Cyber Attack Detection Strategies", *International Journal of Security and Its Applications*, vol. 8, no. 1, pp. 247-256, 2014.

O. Oriola, A. Adeyemo and A. Robert, "Distributed Intrusion Detection System Using P2P Agent Mining Scheme", *African Journal of Computing & ICT*, vol. 5, no. 2, 2012.

R. Hill, "Dealing with cyber security threats: International cooperation, ITU, and WCIT", *2015 7th International Conference on Cyber Conflict: Architectures in Cyberspace*, pp. 119-134, 2015.

Pramod Singh Rathore, "An adaptive method for Edge Preserving Denoising, International Conference on Communication and Electronics Systems, Institute of Electrical and Electronics Engineers & PPG Institute of Technology (2017). *Proceedings of the 2nd International Conference on Communication and Electronics Systems (ICCES 2017)*: 19-20 October, 2017.

P. S. Rathore, A. Chaudhary and B. Singh, "Route planning via facilities in time dependent network," *2013 IEEE Conference on Information & Communication Technologies*, Thuckalay, Tamil Nadu, India, 2013, pp. 652-655. doi: 10.1109/CICT.2013.6558175

N. Bhargava, S. Dayma, A. Kumar and P. Singh, "An approach for classification using simple CART algorithm in WEKA," *2017 11th International Conference on Intelligent Systems and Control (ISCO)*, Coimbatore, 2017, pp. 212-216. doi: 10.1109/ISCO. 2017. 7855983

S. Gordon, R. Ford, (2006) On the definition and classification of cybercrime, *Journal in Computer Virology*, Vol. 2, No. 1, pp. 13-20.

Neeraj Bhargava, Pramod Singh, Abhishek Kumar, Taruna Sharma, Priya Meena, December 17 Volume 3 Issue 12, "An Adaptive Approach for Eigenfaces-based Facial Recognition", *International Journal on Future Revolution in Computer Science & Communication Engineering (IJFRSCE)*, pp. 213–216.

5

Spam Filtering Using AI

Yojna Khandelwal[1]* and Dr. Ritu Bhargava[2]

[1]M.Tech Scholar, MDS University, Ajmer, India
[2]Sohpia Girls' College, Ajmer, India

Abstract

The increase in the number of unwanted mails called spam has created the need for spam filters to reduce time and effort in managing inboxes as well as in managing storage. Efficient spam filters can preclude a user from cyber fraud, and users' data can also be kept secured from spammers. In recent times, machine learning methods have been immensely successful in detecting and filtering spam mails. These models basically "learn" from experience with respect to some tasks and are capable of finding "commonality" in many different observations. This study discusses various methods of spam filtering using existing Artificial Intelligence techniques and compares their strengths and limitations.

Keywords: Inspection, spam, ANN, robot

5.1   Introduction

5.1.1   What is SPAM?

Email or Electronic mail is a very fast as well as economical way of exchanging information over the internet. In the past few decades, use of email has grown immensely and so has the use of illegitimate mails, i.e., spam. Spams are unrequested and undesired emails and are often called as Unsolicited Bulk Mail (UBM); they have became troublesome over the internet. The problem behind spam emails is that they are generally sent in bulk, which leads to wastage of time for filtering a desired set of emails from the inbox as well as locking up a lot of system space and absorbing communication

---

*Corresponding author: yojnakh22@gmail.com

Neeraj Bhargava, Ritu Bhargava, Pramod Singh Rathore, and Rashmi Agrawal (eds.) Artificial Intelligence and Data Mining Approaches in Security Frameworks, (87–100) © 2021 Scrivener Publishing LLC

bandwidth (R. Karthika *et al.*, 2015). Spam filtering is a technique to identify spam mails and keep them separated from ham mails, i.e. useful emails, to reduce time and effort. Knowledge engineering and machine learning techniques have gained some success in spam filtering. On the other hand, the spammers also seek new techniques to bypass the filters, including word obfuscation and image spams.

5.1.2   Purpose of Spamming

Although the user sometimes finds spams irritating, spams have become an integral part of the advertisement industry for businesses. Decades ago, the primary purpose of spammers was to abuse social norms to promote their products but in the current era, the opportunity has taken up by cyber criminals, spam mails are used to collect vital information about the user or sometimes persuade users to visit dicey web links (Singh Rathore, P *et al.*, 2020). Spams, to meet their purpose, must have an intended payload and spam filters must be used in order to protect the user from them.

5.1.3   Spam Filters Inputs and Outputs

As we have discussed, spam filtering is a technique to separate spam mails from ham mails. A filter uses various means to complete the separation task. Sometimes the mail Headers are used to filter while filtering might be on the basis of the content of mails, characteristics of the sender or the target, or maybe the sender to be a spammer.

Figure 5.1 denotes the common word used in the cloud in spam mails. A well-defined set of rules is necessary for a filter to perform the job such as

Figure 5.1 Word cloud of common words in spam mails.

the content of the mails or statistical information derived from the training corpus or feedback methods to apply probabilistic classifiers. The simplest output desired from any spam filter is a binary classification, i.e., a 0 for a sure spam or a 1 for a non-spam. More complex ordinal categorization can include steps such as surely a spam, likely to be a spam, unsure, likely not to be a spam or surely a non-spam.

5.2   Content-Based Spam Filtering Techniques

5.2.1   Previous Likeness–Based Filters

Such filters are generally also called Memory/Instance–based filters. This machine learning–based technique uses the data from the stored mails and classifies the new instance points in accordance with the similarities to the previous stored examples or the training set.

5.2.2   Case-Based Reasoning Filters

Case-based Reasoning filters more often work as reinforcement learning algorithms. In this method the system has a set of previously classified mails/cases, rather than a fixed set of instructions. Every time, at the time of classification, the system tries to match the characteristics of the test case with the former set of cases in the system's collection. Then the test mails will also be included in the system's memory as a classified mail for future

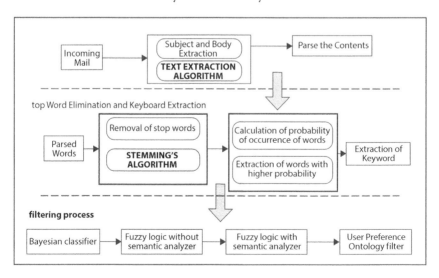

Figure 5.2 A typical spam filter mechanism.

classification. The initial few cases can be classified using Naïve Bayesian classification and then the system will add new ones shown in Figure 5.2 (Sharma and Suryawansi, 2016).

5.2.3   Ontology-Based E-Mail Filters

Ontology-based E-mail spam filtering is a completely different genre of spam detection. This technique was evolved with the use of ontology trees and also with evolution of better classification algorithms. In recent times, the major limitation of this approach was interfacing between two independent systems. To get desired inputs, a prototype is formed that actually uses this information. Also, the major limitation of ontology based filtering was that it is required to pre-process the input mails in CSV format. Along with that it also suffers the limitations with Bayesian classification as it is used to classify the mails for which ontology was used to understand its content (Awad and Elseuofi, 2011; S.P. Rajamohana *et al.*, 2017).

5.2.4   Machine-Learning Models

It is the science of discovering unknowns from the data, obtaining predictive/actionable insights from data, creating data products that have business impacts, communicating relevant business stories from data and building confidence in decisions that drive business values. Machine learning is a program that "learns" from experiences to some task. In simple terms, any machine learning model can be defined as a simplified or partial representation of reality, defined in order to accomplish a task or to reach an argument (P. S. Rathore *et al.*, 2013).

Machine Learning Models include:

5.2.4.1   *Supervised Learning*

Supervised learning is simply a task that is to be done, to extract a description/labelling or pattern from the data, based on the training. In this branch training examples are labelled by a supervisor (human) that in turn is used to predict the output for further examples. Some common applications of supervised learning include credit approval, medical diagnosis and fraud detection, etc. (N. Bhargava *et al.*, 2017).

5.2.4.2   *Unsupervised Learning*

Unsupervised learning is basically finding interesting patterns/groups/categories in the data based on evidence. In this branch of machine learning there is no relabeled data used. Performance of how good the group/pattern is measured from raw data (Naveen Kumar *et al.*, 2018). Some common application of unsupervised learning are customers segmentation, user behavior categorization and grouping of items by similarities, etc.

5.2.4.3 *Reinforcement Learning*

Reinforcement learning is a separate arm of Machine Learning, that is sometimes also known as outline learning. These machine learning models basically learn through trial and error interactions. The goal is to maximize the amount of reward received from the environment. It is an interactive process in which a model is trained by interaction through the environment and a reward is assigned on every success. The performance is measured in terms of amount of rewards received. Some other common applications are Robot learning and games, etc. (Bharat Singh *et al.*, 2013).

5.3 Machine Learning–Based Filtering

5.3.1 Linear Classifiers

A linear classifier is used to classify say n features represented in matrix form as say x[m] for any point m in a n dimensional space. Here the coefficient matrix is defined as $\beta = (\beta 1 \beta 2 \cdots \beta n)$ with threshold t, then the hyper plane that divides the plane into two halves will follow the equation $\beta \cdot x = t$. Hence all the points lying above the plane, i.e., with $(\beta \cdot x[m] > t)$ are termed as spam and others as non-spam.

Algorithm – To intuitively find a separating hyper plane, if one exists, the weights can be increased or decreased for every point on the wrong side. Here the algorithm ignores the correctly classified examples. Now the training can be stopped under the assumption that a sufficiently good classifier is found over the training set. This method is quite simple, efficient as well as incremental. Here adding rate and margin parameters can also be used to effect training on near-misses and errors shown in Figure 5.3 (N. Bhargava *et al.*, 2017).

Vector representation of some messages in training set of our running example. Here the separating hyperplane sets perfectly for the training data.

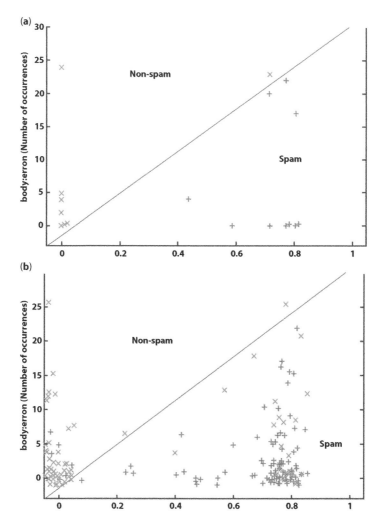

Figure 5.3 Hyperplane sets for tanning data.

Here the same hyperplane obtained above results to fail when used of the whole sample data. The hyperplane is able to keep most of the spam mails on the spam side and non-spam on the other. So this can be considered a reasonable classifier but it is doubtful that it is the best.

5.3.2 Naïve Bayes Filtering

Naïve Bayes is also a supervised learning technique based on probability and statics. This method of filtering mails uses an adaptive set of rules and the associated set of probabilities are set according to the classification decisions and mails received. Each mail is described by a set of attributes and each attribute is assigned a probability according to the number of times it has occurred in the training set.

Naïve Bayes classifier for filtering spams uses a simple probability formula which can be interpreted as (where c=spam): "The probability of a message being spam, given its features equals the probability of any message being spam multiplied by the probability of the features co-occurring in a spam divided by the probability of observing the features in any message" shown in Figure 5.4.

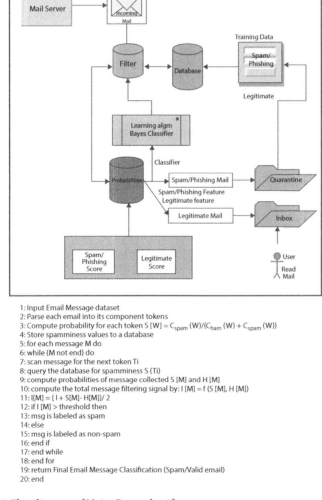

1: Input Email Message dataset
2: Parse each email into its component tokens
3: Compute probability for each token S [W] = C_{spam} (W)/(C_{ham} (W) + C_{spam} (W))
4: Store spamminess values to a database
5: for each message M do
6: while (M not end) do
7: scan message for the next token Ti
8: query the database for spamminess S (Ti)
9: compute probabilities of message collected S [M] and H [M]
10: compute the total message filtering signal by: I [M] = f (S [M], H [M])
11: I[M] = {1 + S[M]- H[M]}/ 2
12: if I [M] > threshold then
13: msg is labeled as spam
14: else
15: msg is labeled as non-spam
16: end if
17: end while
18: end for
19: return Final Email Message Classification (Spam/Valid email)
20: end

Figure 5.4 Flowdiagram of Naive Bayes classifier.

5.3.3   Support Vector Machines

Support Vector Machines is a supervised learning algorithm that has shown quite better performance than other classifiers because of its multidimensional boundaries and simplicity. It maximizes the distance to the nearest example point and the points that are equidistant to the given example point are termed as support vectors. A linear combination of these support vectors forms a classifier or a separating hyperplane.

Algorithm: Input a training set, say S, and define a kernel function of form {c1, c2,…..,cn} and {d1, d2,…..,dn}. Assign a number of nearest neighbors, say k. Then design a two stage for loop, for the outer loop set c=ci from 1 to n. The inner loop holds for j from 1 to q, where a SVM classifier function f(x) is designed with merger parameters (c, d). Using an if-else condition the classifier function f(x) is compared with the best classifier given by k-fold cross validator. Hence a return command is given to classify a message as spam or a non-spam shown in Figure 5.5 (S K Rajesh Kanna *et al.*, 2014).

5.3.4   Neural Networks and Fuzzy Logics–Based Filtering

Neural networks are a group of simple processing units based on biological neural network. Each unit is connected with its neighbors with some assigned weights. Each of them accepts the input from one neighbor and transmits the modified output to the other neighbor. More generally a neural network is designed by interconnecting the three layers, i.e., the input layer, the middle layer and the output layer.

Fuzzy logic is a also a methodical approach to classify mails as spam or legitimate automatically by taking the content of the mail into consideration.

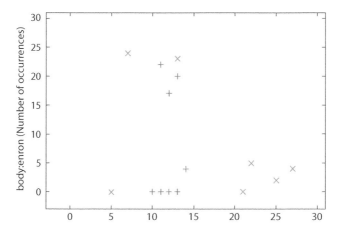

Figure 5.5 Simulation results of a SVM classifier.

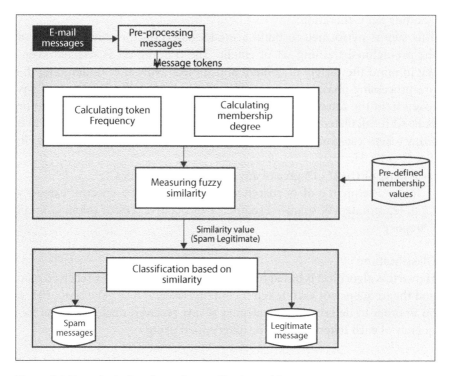

Figure 5.6 Fuzzy logic–based e-mail spam filtering architecture.

The program can automatically adapt from the mails' content and build its own database. A classifier will be built from a training set of pre-classified mails (S K Rajesh Kanna *et al.*, 2017).

Email spam filtering system architecture is built in three steps, i.e., pre-processing, training set generation and classification shown in Figure 5.6.

Pre-processing:
Data pre-processing is the most important step before applying any logic to the training set. Emails also need to be pre-processed before using them for training and classification for making the algorithm more efficient as well as for obtaining optimum results. This step includes cleaning the texts by removing all HTML tags along with getting rid of non-significant words such as punctuation, etc. Then stemming is done to ultimately reduce the sparsity of the final matrix of words of the form T = <t1, t2… tN> where N is the total number of significant words that can now be termed as tokens. Now the number of times each token has occurred in each category, c belongs to spam, legitimate}, is determined (Pramod Singh Rathore, 2017).

Training Set Generation:

This step is performed to build a model based on the characteristics of the pre-defined training set of emails. The training set is selected keeping in mind the variety of content and subject. Now after undergoing the pre-processing phase we can get (fi,c) which denotes the frequency of any token ti, in the category 'c'. From this set of data a neural network can be built with calculated weights of each token to form matrix T x C and that fuzzy token- category relation is used to assign the value of relation, i.e., R: $T \times C \rightarrow [0, 1]$.

Where, $\mu R(ti,cj)$ - Degree of any token ti, in category Cj.

fi cj - Frequency of occurrences of token ti in a specific category Cj fi, legitimate +fi, spam - Frequency of occurrences of token ti in all Category

Classification:

Hence this algorithm is based on fuzzy relations between any received mail and the frequency of each token it contains (Neeraj Bhargava *et al.*, 2017). So in order to determine the category of any received mail, say d, the frequency of each token in d can be determined using -

$$\mu_d(t_i) = \frac{f_{i,d}}{\max_{t_j \in d}\{f_{j,d}\}}$$

Then fuzzy conjunction and fuzzy disjunction operators can be used to measure the ratio of similarity and the threshold value to compute the fuzzy similarity shown in Figure 5.7.

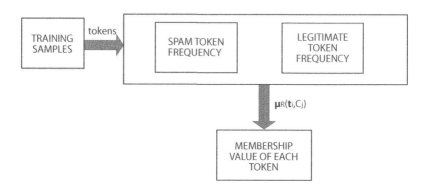

Figure 5.7 Block diagram for training set generation.

5.4   Performance Analysis

A sample corpus of emails can be taken and incorporated to Header Based Filtering, Content-Based Filtering or any URL filtering technique. Here also we are considering a sample set of emails and having gone through different classifiers/filtering methodologies the following quantitative comparison is noted in terms of model's Accuracy rate, recall rate and precision value.

a. Ontology-based E-mail filter - In around 5-6 trials, the accuracy of this model varied between 0.95 to 1 which showed that increasing the number of inputs leads to fall in accuracy. Its recall value fluctuates from 0.94 to 1 while precision value lied between 0.97 to 1.

b. Linear Classifier – A different dataset is used for the classification for this technique. The accuracy comes out to be 97.72% using the algorithm described above. Its recall rate was noted as 97.29% and the precision as 97.55%.

c. Naïve Bayes Filtering – This model has quite a lesser accuracy, i.e., between
0.78 to 0.8 as increasing the number of inputs needs more training. Recall lies in range 0.85 to 0.91 and precision fluctuates from 0.77 to 0.91 showing the insufficient knowledge base is used to train the model.

d. Support Vector Machines – This classifier has an accuracy rate of 0.96 showing the classifier is quite good to use for filtering. Its recall rate comes out to be 95% and precision value as 93.12%.

e. Neural networks and fuzzy logics–based filtering – Again the accuracy rate for this model ranges from 0.8 to 0.9 on different trails over a dataset showing the fall in accuracy value due to increase in number of inputs because of insufficient training. Its recall rate varied from 0.87 to 0.91 while precision ranges from 0.84 to 0.97.

5.5   Conclusion

In the discussion above, we have reviewed some of the most popular and coherent machine learning algorithms used to solve the problem of spam emails. The study makes a clear performance analysis and comparison of various content-based spam filtering techniques. The experiment shows a quantitative comparison of the listed techniques in terms of Accuracy, Recall and Precision values.

Most of the machine learning models are built on the training data and hence their accuracy varies. Increasing the number of inputs needs more training, i.e., a large and sufficient knowledge base containing most of the variety of data. Larger test sets show lesser accuracy because of insufficient data set during the training phase. Linear classifier has shown a good accuracy but is more specific to the input datasets while more generally support vector machines have shown a satisfactory performance. Bayesian's approach in email spam detection has provided a fine base for creating a Meta-spam classifier. More research is needed on neural networks and fuzzy logics–based system to find a much more efficient way for spam filtering not only in terms of content-based filtering but also spam image detections.

References

R. Karthika, P. Visalakshi, A hybrid ACO based feature selection method for email spam classification, *WSEAS Trans. Comput.* 14 (2015) 171–177.

Singh Rathore, P., Kumar, A., & Gracia-Diaz, V. (2020). A Holistic Methodology for Improved RFID Network Lifetime by Advanced Cluster Head Selection using Dragonfly Algorithm. *International Journal of Interactive Multimedia and Artificial Intelligence*, 6 (Regular Issue), 8. http://doi.org/10.9781/ijimai.2020.05.003

A. Sharma, A. Suryawansi, A novel method for detecting spam email using KNN classification with spearman correlation as distance measure, *Int. J. Comput. Appl.* 136 (6) (2016) 28–34.

W.A. Awad, S.M. Elseuofi, Machine learning methods for spam E-mail classification, *Int. J. Comput. Sci. Inf. Technol.* 3 (1) (2011) 173–184.

S.P. Rajamohana, K. Umamaheswari, B. Abirami, Adaptive binary flower pollination algorithm for feature selection in review spam detection, in: *IEEE International Conference on Innovations in Green Energy and Healthcare Technologies*, 2017, pp. 1–4.

P. S. Rathore, A. Chaudhary and B. Singh, "Route planning via facilities in time dependent network," *2013 IEEE Conference on Information & Communication Technologies*, Thuckalay, Tamil Nadu, India, 2013, pp. 652-655. doi: 10.1109/CICT.2013.6558175

N. Bhargava, S. Dayma, A. Kumar and P. Singh, "An approach for classification using simple CART algorithm in WEKA," *2017 11th International Conference on Intelligent Systems and Control (ISCO)*, Coimbatore, 2017, pp. 212–216. doi: 10.1109/ISCO. 2017. 7855983

Naveen Kumar, Prakarti Triwedi, Pramod Singh Rathore, "An Adaptive Approach for image adaptive watermarking using Elliptical curve cryptography (ECC)", *First International Conference on Information Technology and Knowledge Management* pp. 89–92, ISSN 2300-5963 ACSIS, Vol. 14 DOI: 10.15439/2018KM19

Bharat Singh, Ravinder Singh and Pramod Singh Rathore. Article: Randomized Virtual Scanning Technique for Road Network. *International Journal of Computer Applications* 77(16):1–4, September 2013.

Prof. Neeraj Bhargava, Abhishek Kumar, Pramod Singh, Manju Payal, December 17 Volume 3 Issue 12, "An Adaptive Analysis of Different Methodology for Face Recognition Algorithm", *International Journal on Future Revolution in Computer Science & Communication Engineering (IJFRSCE)*, pp. 209–212

N. Bhargava, S. Sharma, R. Purohit and P. S. Rathore, "Prediction of recurrence cancer using J48 algorithm," *2017 2nd International Conference on Communication and Electronics Systems (ICCES)*, Coimbatore, 2017, pp. 386–390, doi: 10.1109/CESYS.2017.8321306.

S K Rajesh Kanna, G Muthu and A. Venkatesan, "Inspection of Boiler Pipes Using Miniature Mobile Robot", *International Journal of Advanced Research in Management, Engineering and Technology*, 2017, vol. 2, issue 4, pp. 677–680.

Pramod Singh Rathore, "An adaptive method for Edge Preserving Denoising," International Conference on Communication and Electronics Systems, Institute of Electrical and Electronics Engineers & PPG Institute of Technology. (2017). *Proceedings of the 2nd International Conference on Communication and Electronics Systems (ICCES 2017)*: 19–20, October 2017.

Neeraj Bhargava, Pramod Singh, Abhishek Kumar, Taruna Sharma, Priya Meena, December 17 Volume 3 Issue 12, "An Adaptive Approach for Eigenfaces - based Facial Recognition", *International Journal on Future Revolution in Computer Science & Communication Engineering (IJFRSCE)*, pp. 213–216.

6

Artificial Intelligence in the Cyber Security Environment

Jaya Jain

M.tech Scholar, MDS University, Ajmer India

Abstract

AI knowledge is presented by machines. Any program that observes its trend and finds a way to take steps to increase its chances of accomplishing a task can be described as AI. The study of AI is rich and novel, such as PC understanding, such as, deep learning, AI, normal language preparing, thinking and settling on dynamic innovation, discourse and survey innovation, human interface innovation, semantic innovation, exchange and story handling, among different advances.

The aptitude of the top specialists is made promptly accessible to all, where their innovation is enacted through an understanding system. With repeated use, this framework will give more accurate answers, eventually surpassing the exactness of human specialists.

As machine intelligence improves, machines will figure out how to comprehend the incorporated information on humankind. With the utilization of computerized sensor information, intelligent-based gadgets can be utilized to create smart consultants, instructors or collaborators.

As artificial intelligence advances, there are risks associated with their utilization, set up in functioning frameworks, tools, calculations, framework the executives, morals and duty, and privacy. This study focuses on the risks and threats of computerized reasoning and how AI can help comprehend network safety or areas of cyber security issues.

Keywords: Inconsistency, artificial intellect, intellectual abilities, network safety/cyber security, supervised machine learning, unread machine learning

Email: jayavineetjain@gmail.com

Neeraj Bhargava, Ritu Bhargava, Pramod Singh Rathore, and Rashmi Agrawal (eds.) Artificial Intelligence and Data Mining Approaches in Security Frameworks, (101–118) © 2021 Scrivener Publishing LLC

6.1 Introduction

Initially, Artificial Intelligence (AI) was an idea to mirror the human cerebrum, and to explore real-world issues in a holistic manner. Computer-based intelligence is most popular for its film and scholarly applications; AI makes it conceivable to supply a lot of information and to deal with that information wisely [1]. Artificial intelligence has been utilized to give intelligent applications in an assortment of zones, for example, safeguard or space investigation. These destinations have a rich history of different troubleshooting arrangements. Later, AI saw application in the field of medical services. It has been utilized for things like diagnostics, treatment suggestions, and careful treatment [2].

Performance technology can be characterized as artificial intellect that provides tools to solve complex and stressful issues. Artificial intelligence is a combination of data innovation and physical intelligence, which can be utilized electronically to accomplish goals. Wisdom is the capacity to reflect by building recollections and accepting, seeing examples, assembling powerful selections, and gaining for a fact. AI can make machines perform like humans, but they can be quicker and additionally more gentle [3].

The majority of these gatherings at the security level control the safety areas.

The arrangement standard utilized in this investigation can be perceived as scientific classification [4]. We are introducing some areas of cyber security shown in Figure 6.1:

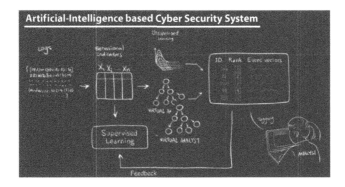

Figure 6.1 Cyber security system.

- Structure safety
- End safety
- Application safety
- IoT safety
- Web safety
- Safety and event response
- Intelligence threats
- Mobile safety
- Cloud safety
- Ownership and access management
- System safety
- Manpower safety

This investigation examines presentations that utilize man-made reasoning from various makers. The applications being considered utilize man-made consciousness to foresee and distinguish data security dangers and failures. The conversation of the applications intends to give a diagram of network safety arrangements utilizing accessible artificial intelligence, and what they can offer to determine the solution to the problem [5].

Cyber security

Digital protection measures are related with hazard the executives, hazard catch and framework vigor. Key exploration themes incorporate procedures identified with recognition of noxious organization conduct and malware, just as IT inquiries identified with IT security. So, network protection can be characterized as the different activities taken to forestall digital assaults and their outcomes and incorporate the utilization of essential enemy of debasement measures. Network protection is based on the threat examination of a link or establishment. The construction and parts of an association's network safety procedure and its execution plan depend on quantifiable dangers and danger investigation. The purpose is to set up various authoritative security procedures and rules [6].

Threats that are are directly or indirectly targeted at public or private projects emerge from inside or outside national borders. A threats circumstance is an angle of threats that contain data about mishap threats and attack ships. Through misuse or shortcomings, threats lead to misfortune or reallocation of property [7].

A significant factor is the fundamental planning to combat threats, and sufficient protection against the harmful impacts of threats. Courses of

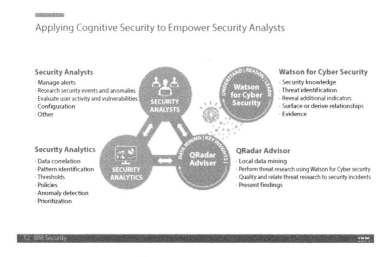

Figure 6.2 Empower security analysts.

action to battle digital threats can be improved by improving the essentials of network safety, expanding everybody's information about danger, improving execution and maintenance of security [8]. The key is to recognize network safety challenges and react appropriately. A significant piece of network protection is the ability to hold the capacity to work under a digital assault, to have the choice to rapidly end an attack and rebuild the link's tasks to its typical pre-occasion state. Suitable laws and proper, in-depth talk are needed to resolve these issues. Ways to combat cyber-attacks can be discussed in detail shown in Figure 6.2 [9].

6.2   Digital Protection and Security Correspondences Arrangements

Artificial intelligence not only poses threats and dangers; it can also serve as a solution to problems. Information passage and operational direction are utilized to recognize, forestall and identify cyber-attacks. Daily data safety schedules are man-made or mechanized. These so-called analytical diagnostic activities depend on instructions made by IT security experts, who ignore attacks that are not in accordance with set up rules. Mechanically put together strategies can bring about false positive outcomes, create a common sense of distrust of the system and need human effort to examine cases [11].

6.2.1   Operation Safety and Event Response

Man-made consciousness is one suitable choice that can conceivably fore-stall huge loss of lives as ... underneath made by Eastern Kentucky College's Online Bosses in Security degree program. ... Man-made brainpower for Calamity Reaction (AIDR) ... picture during crisis activities to be utilized by crisis activity focuses.

6.2.2   AI2

Proceedings of the Twelfth Global Meeting on Military Safety and Security ICCWS2019
IMIT Computer Discipline and the Artificial Intelligence Laboratory (CSAIL) and Patrice have made an AI2 spy stage to foresee digital assaults. As per Conner-Simons, the AI2 stage had the option to accomplish 86% exactness in identifying digital assaults, which was multiple times superior to past examinations. To prevent attacks AI2 recognizes dubious action utilizing information section calculations utilizing AI calculations. Prevent attacks Investigators likewise added to the stage models (directed readings) in the accompanying informational collection, which empowers further learning [12, 13].

The program is also ready to continuously create new representations within hours, which could essentially improve the speed of its digital assault abilities.

6.2.2.1   CylanceProtect

CylanceProtect is an incorporated data safety tool, which joins the advantages of man-made brainpower and data safety panels to avoid malware contaminations. Data security panels are utilized to ensure against script attacks, memory assaults or outside gadgets. Dissimilar to traditional security gears dependent on investigation examination and client conduct in identifying natural dangers, CylanceProtect:

- Uses artificial brainpower (not signatures) to recognize and secure well-known and unknown malicious software running on end devices
- Avoids known and obscure zero-day assaults
- Keeps gadgets without upsetting the end client

6.3   Black Tracking

Figure 6.3 AI cyber security.

The dark track is a data safety organization, which can help distinguish and identify developing digital dangers that could prevent the assurance of conventional data. Dark trace utilizes Enterprise Immune System (EIS) innovation and utilizes AI calculations and numerical standards to identify negligence inside the association's data organization. The EIS uses statistical techniques, which implies it doesn't have to exploit marks or leads, and can highlight uncommon digital protection assaults. The EIS can recognize and react to some all-around planned digital dangers, incorporating inward dangers covered up in data organizations. By utilizing learning and measurements instruments, the EIS can adjust and subsequently figure out how every client, gadget and data network acts, so as to distinguish conduct that reflects genuine digital dangers. Dim exploration innovation gives organizations an all-encompassing perspective on the data organization and permits them to react all the more unequivocally to dangers and lessen hazard shown in Figure 6.3 [16].

It views the info system and permits it to reply more strongly to threats and decrease risk. Instead of describing "bad" behaviors in the past and relying on previous attack methods, dark trace machine reading, and a Bayesian opportunity perspective, can automatically model and integrate data with power and speed. The black track monitors raw data, such as cloud service integration, which is transmitted over the network in actual time, without interruption, such as, business operations and connections. It also delivers a straight opinion of all digital events by writing continuous or malicious assaults.

The black track involves four mathematical machines, which use numerous mathematical methods, like the duplication of the Bayesian equation.

The first three models produce communication models for individuals and the devices they use, as well as groups as a whole. When abnormal performance is detected, one or more of these engines show a message to the danger separator, whose function is to detect serious circumstances and malfunction reports that cannot be properly examined. The grouping of consistent Basest approaches makes for the exact accuracy of the disruption on the organizational scale [17].

6.3.1   Web Security

Cyberlytic Profile:

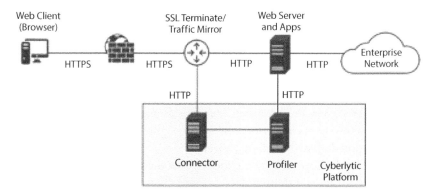

Figure 6.4 Cyberlytic profile.

Cyberlytic Profiler utilizes an assortment of novel procedures to dissect capricious outcomes from web traffic to decrease the accessibility of standard marks and authority innovation. These techniques help distinguish digital dangers that are continually expanding and complex by decreasing the requirement for human mediation or coordinated effort. Profiler utilizes an electronic and arbitrary AI program to investigate information streams. Free self-study calculations produce a standard web conduct by settling on choices all alone. Accessible headings fluctuate from web worker and incorporate occasional styles and techniques. Field highlights incorporate length and succession appropriation of characters shown in Figure 6.4 [9].

By printing web applications, Profiler can decide if the introduced applications are from the ordinary of distribution the application in a specific region of the web application. This technique figures out what is the "normal" in a particular group.

As a decisions can be attracted to decide the motive, making the traffic disliked. By highlighting weaknesses, the most probable disturbances can be stressed; unlawful can be evaluated for hazard. Profiler utilizes an alternate protected technique to distinguish assault highlights of these kinds of assaults: SQL infusion, compose site (XSS) and Bash. (Cyberlytic.)

6.3.1.1   Amazon Macie

Amazon Macie is an info safety facility that utilizises AI. The artificial intelligence consciousness gives Macie identification tools to identify, arrange and shield delicate data from Amazon Web Services (AWS). Macie views delicate records as special info or copyright Likewise, it can observe how copyrighted material, for example, documents, are copied, distributed, or viewed. Macie has a dashboard view which decides how information is utilized or erased. The application keeps on observing information utilization and inconstancy and gives itemized alerts or warnings if data is subject to unauthorized use or unintended document leaks shown in Figure 6.5 [10].

Amazon Macie is a data security service that uses AI. The man-made brainpower consciousness gives Macie distinguishing proof devices to recognize, mastermind and shield delicate information from Amazon Web Services (AWS). Macie views delicate data as particular information or copyright. Likewise, it can observe how copyrighted material, for example, forms, are copied, distributed, or viewed.

Macie has a dashboard which chooses how data is used or deleted. The application continues watching data use and changeability and gives separated alarms or alerts if information is dependent upon unapproved use or unintended information spills.

Macie can also repeatedly identify the danger of business information, if the information is distributed external to the group without permission, or if such data is inadvertently accessed. Macie can likewise consequently identify the danger of business data, if the data is appropriated outside the association without approval, or if such information is unintentionally accessed shown in Figure 6.5 [11].

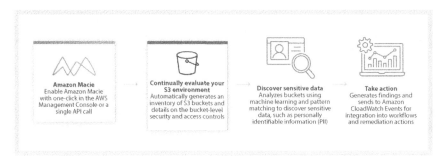

Figure 6.5 Amazon Macie.

Deep sensitivity software is designed to *keep* cell phones and service area from well-known and unknown malicious assaults in actual time; the product depends on the misuse of fake neural organizations. With the installation technology, Deep Instinct can recognize software on cell phones, and services workstations. Utilizing the appropriate in-depth reading skills, the software is capable to anticipate anonymous cyber-attacks shown in Figure 6.6 and Figure 6.7 [12].

The designers of Deep sensitivity have used progressed learning calculations in their application of the system, enabling them to detect structures used in malicious software. Deep Instinct can identify and forestall noxious programming improvement at all degrees of the association. Utilizing inside and outside understanding abilities, Deep Instinct architects constructed a thorough neural organization in lab conditions and trained it with an enormous arrangement of malicious code tests. The data was utilized with dangerous and harmless documents to show these neural organizations. The outcome was a prescient model, which could be sent to the device for assurance to give ongoing support [13].

The knowledge consists of getting software to recognize a combination of applications and applications based on malicious software. The Deep Instinct learning method cuts software code examples with very small captions, to test malicious software.

Proceedings of the Twelfth Global Meeting on Military Safety and Safety ICCWS2019

Figure 6.6 Deep sensitivity.

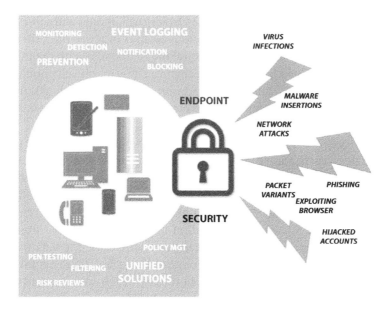

Figure 6.7 ENDPOINT security.

This technique is like the genomic sequence, in that it similarly contains tens of thousands of slighter series. These sample fragments are inserted into neural networks to show the network for targeting resolves. This type of system executes a big, difficult calculation, and a GPU set is used to assist with these calculations. GPU calculation power is much faster than CPUs. The outcome is a fast and mathematical neural network that involves slight computer control, and can be used to detect malicious software.

6.4   Spark Cognition Deep Military

Spark Cognition Deep Armor is capable to identify and protect against the risk of malware, worms, viruses Trojans and viruses lists, utilizing numerical strategies, for example, AI and normal language correction. The Deep Armor configuration incorporates a little start to finish operator coordinated with a cloud-based engine, as well as a threat detection policy. The end operator recognizes and hinders noxious projects and other significant-level dangers, notwithstanding marks. The specialist is intended to ensure customers, workers, portable and IoT gadgets; Spark, the cloud-based comprehension engine for Deep Armor explanation uses new level filters shown in Figure 6.8 [14].

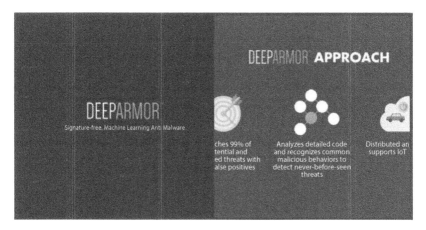

Figure 6.8 Cloud-based comprehension engine.

6.5   The Process of Detecting Threats

The main layer of security performs document investigation, just as the use and risk management board, to rapidly recognize known or new files. After filtering the recovered files records, Deep Armor utilizes intellectual calculations to examine obscure documents and threat forms for each file. In the subsequent stage, recognizing the threat and the cloud-based support gives a natural language handling tool (NLP). Deep NLP not only understands the online proof, but also understands the context surrounding the threats; Deep Armor can recognize perilous materials as indicated by abnormal cases [18].

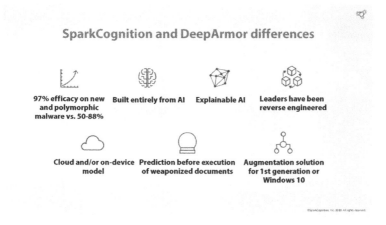

Figure 6.9 The process of detecting threats.

SparkCognition's Deep NLP innovation works so that a great many pages of pertinent data identified with different threats are considered tools to incorporate Spin Cognition's Deep NLP technology. This information is utilized to distinguish the threat itself. NLP tools also test the Internet for proof of dangers, from which a summary of evidence will be created. The goals of the risk calculation will also be determined based on recognized risk factors. Finally, a synopsis of the risk analysis can be created from the produced data that can be used to plan strategy procedures and address the most applicable issues shown in Figure 6.9 [15].

6.6   Vectra Cognito Networks

Vectra Cognito Networks utilizes man-made consciousness to deliver an itemized picture of an ongoing digital assault. Cognito joins refined mechanical learning advances, for example, top to bottom learning and neural organizations, and constant learning models, to rapidly and successfully recognize covered up and obscure assailants before they cause damage. Cognito additionally removes removes supposed "blind areas" by dissecting all info safety and verification frameworks and SaaS applications for network traffic and log documents. This gives a total diagram of the status of clients and IoT gadgets identified with measures working in cloud and server farms, keeping assailants from covering up shown shown in Figure 6.10 [19].

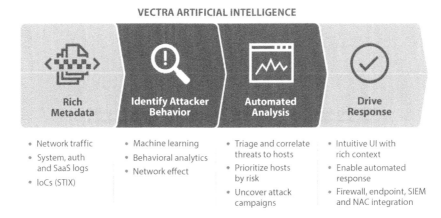

Figure 6.10 Vectra AI.

Proceedings of the Twelfth Global Meeting on Military Safety and Security ICCWS2019

Vectra Cognito utilizes observation and uncontrolled electronic learning innovations, for example, top to bottom learning and neural organizations, to battle digital assaults and target them. Indigenous data security frameworks endeavor to distinguish digital assaults via looking for marks definitely known and misused. A digital assailant can utilize this data against the framework. Cognito learns network action over some undefined time frame, for example, days, weeks, or months [20].

Cognito recognizes the conduct of digital assailants in each progression of digital assaults. Observed aggressors' conduct is ordered and contrasted with typical client conduct, utilizing workers that have just been tried for hazard. The individuals who are essential for each composed digital assault crusade are distinguished as mirroring the conduct of the aggressors. For this situation, chiefs can zero in on guiding their assets to the most risky assaults on business.

Risks of threats

IBM QRadar Counsellor with Watson utilizes IBM Watson's smart talents (i.e., computerized reasoning) stage QRadar Safety, a stage intended for data safety examination, uncovering concealed dangers and making check dangers. The framework naturally identifies hazard pointers, utilizes knowledge to utilize its psychological capacities to get delicate data and at last quickens the pattern of reaction to security dangers. Qonad Advisor with Watson likewise utilizes Watson's Cyber Safety highlights to research and react to data safety dangers [21].

Figure 6.11 QRadar Advisor.

The QRadar Counsellor with Watson works through the accompanying advances shown in Figure 6.11:

- When the Q Radar Safety Intelligence stage distinguishes security data dangers, a data security expert may allude it toQ Radar Advisor and Watson for additional examination. Guide starts a total sweep of security data dangers with mining information from nearby QRadar programming. The product will at that point utilize Watson's Cyber Security to lead an intensive investigation of the danger [18].
- Watson's Cybersaftey gathers information from an assortment of sources, for example, sites, data security gatherings and news releases, in a way that is meaningful to the person. Last, the product requires extra data about security data identified with pernicious documents and dubious IP addresses [22].
- Finally, QRadar and Watson butt-centric counsel

User Behavior Analytics (UBA) has been gaining a lot of attention in the data security domain. While establishing corporate security against outside dangers, associations shield themselves from expected dangers inside the association. Dangers can be made, for instance, by a representative or an external player, who can cause harm in a specific region due to negligent behavior. Such dangers are a test of character and can genuinely harm the association's organization's assets, weaken its intangible assets and consumer confidence and damage the reputation or reputation of the organization shown in Figure 6.12 and Figure 6.13 [23].

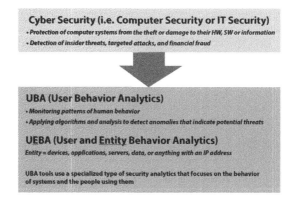

Figure 6.12 Cyber Security/UBA/UEBA.

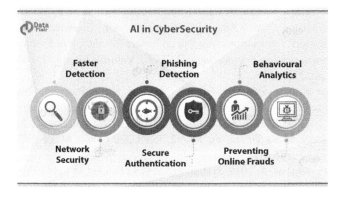

Figure 6.13 AI in CyberSecurity.

6.7 Conclusion

The chapter identified the 12 important areas of cyber safety. During the study, data were collected on 11 artificial intelligence solutions. These results are divided into the areas: Structure safety, end-to-end safety, web safety, security and accountability, spy threats, mobile security and manpower security. The solution to the problem lies in the widespread distribution of cyber security threats.

References

1. Amazon Web Services, Inc. 2018. Amazon Macie FAQ. Amazon. Retrieved 17.10.2018 https://aws.amazon.com/macie/faq
2. Buczkowski, A. 2017. What's the Difference Between Artificial Intelligence, Machine Learning and Deep Learning? GEO. Retrieved 31.5.2017
3. Singh Rathore, P., Kumar, A., & Gracia-Diaz, V. (2020). A Holistic Methodology for Improved RFID Network Lifetime by Advanced Cluster Head Selection using Dragonfly Algorithm. *International Journal of Interactive Multimedia and Artificial Intelligence*, 6 (Regular Issue), 8. http://doi.org/10.9781/ijimai.2020.05.003
4. P. S. Rathore, A. Chaudhary and B. Singh, "Route planning via facilities in time dependent network," *2013 IEEE Conference on Information & Communication Technologies*, Thuckalay, TamilNadu, India, 2013, pp. 652-655. doi: 10.1109/CICT.2013.6558175
5. Bharat Singh, Ravinder Singh and Pramod Singh Rathore. Randomized Virtual Scanning Technique for Road Network. *International Journal of Computer Applications* 77(16):1-4, September 2013.

6. Prof. Neeraj Bhargava, Pramod Singh, Abhishek Kumar, Taruna Sharma, Priya Meena, "An Adaptive Approach for Eigenfaces-based Facial Recognition", *International Journal on Future Revolution in Computer Science & Communication Engineering (IJFRSCE)*, Volume 3 Issue 12, December 17, 2017, pp. 213 – 216.

7. Naveen Kumar, Prakarti Triwedi, Pramod Singh Rathore, "An Adaptive Approach for image adaptive watermarking using Elliptical curve cryptography (ECC)", *First International Conference on Information Technology and Knowledge Management*, pp. 89–92, ISSN 2300-5963 ACSIS, Vol. 14 DOI: 10.15439/2018KM19

8. Conner-Simons. A. 2016. System Predicts 85 Percent of Cyber-Attacks Using Input from Human Experts.Science X network. Retrieved from https://phys.org/news/2016-04-percent-cyber-attacks-human-experts.html

9. N. Bhargava, S. Dayma, A. Kumar and P. Singh, "An approach for classification using simple CART algorithm in WEKA," *2017 11th International Conference on Intelligent Systems and Control (ISCO)*, Coimbatore, 2017, pp. 212-216. doi: 10.1109/ISCO.2017.7855983

10. Cyberlytic. The Profiler – AI for Web Security Technical Data Sheet. Cyberlytic. Retrieved 17.10.2018 https://www.cyberlytic.com/uploads/resources/Technical-Data-Sheet-Final.pdf

11. Pramod Singh Rathore, "An adaptive method for Edge Preserving Denoising," International Conference on Communication and Electronics Systems, Institute of Electrical and Electronics Engineers & PPG Institute of Technology (2017). *Proceedings of the 2nd International Conference on Communication and Electronics Systems (ICCES 2017)*: 19-20 October 2017.

12. Darktrace. 2018. Darktrace Enterprise - Detects and classifies cyber-threats across your entire enterprise. Darktrace. Retrieved 17.10.2018 https://www.darktrace.com/en/products

13. FinancesOnline. 2018. Financesonline IBM MaaS360 Review. Retrieved 22.10.2018 https://reviews.financesonline.com/p/ibm-maas360

14. GeeksforGeeks. Artificial Intelligence | Natural Language Generation. A Computer Science Portal for Geeks. Retrieved 30.3.2018 https://www.geeksforgeeks.org/artificial-intelligence-natural-language-generation

15. IBM QRadar Advisor with Watson. What can Artificial Intelligence Do for Security Analysis? IBM QRadar Advisor with Watson. Retrieved 20.10.2018 https://www.ibm.com/us-en/marketplace/cognitive-security-analytics

16. Kannan, P. V. 2017 Artificial Intelligence – Applications in Healthcare. Asian Hospital & Healthcare Management. Retrieved 30.5.2017 https://www.asianhhm.com/technology-equipment/artificial-intelligence\

17. Kasperskylab. 2018. What is Cyber-Security?AO Kaspersky Lab. Retrieved 8.1.2018 https://www.kaspersky.com/resource-center/definitions/what-is-cyber-security

18. Lehto, M. 2015. Phenomena in the Cyber World. In M. Lehto & P. Neittaanmäki (Eds.) *Cyber Security: Analytics, Technology and Automation*. Berlin: Springer.

19. Lord, N. 2017. What is Cyber Security? Data Insider. Retrieved 8.1.2018 https://digitalguardian.com/blog/what-cyber-security

20. Palmer, T. 2017. Vectra Cognito – Automating Security Operations with AI. ESG Lab Review. Retrieved 21.10.2018 https://info.vectra.ai/hs-fs/hub/388196/file-1918923738.pdf

21. Patel, M. 2017. QRadar UBA App Adds Machine Learning and Peer Group Analyses to Detect Anomalies in User's Activities. IBM. Retrieved 23.10.2018 https://securityintelligence.com/qradar-uba-app-adds-machine-learning-and-peer-group-analyses-to-detect-anomalies-in-users-activities

22. Selden, H. 2016. Deep Instinct: A New Way to Prevent Malware, with Deep Learning. Tom's hardware. Retrieved 18.10.2018 https://www.tomshardware.com/news/deep-instinct-deep-learning-malware-detection,31079.html

23. SparkCognition. 2018. A Cognitive Approach to System Protection. SparkCognition. Retrieved 18.10.2018 https://www.sparkcognition.com/deep-armor-cognitive-anti-malware

7

Privacy in Multi-Tenancy Frameworks Using AI

Shweta Solanki

M.tech Scholar, MDS University, Ajmer, India

Abstract

Multi-tenancy technology is very popular, and development of this concept is also increasing rapidly. The tenants share software and hardware. But the concept of privacy also has a main role in this system since each tenant has access only to their own data. But all tenants' data is stored in the single or one database. So privacy and security is the main problem in the multi-tenancy framework. For that problem we use the artificial intelligence concept to improve the security and privacy concept in multi-tenant-based system.

Using Artificial intelligence makes the privacy and security concept strong because in artificial intelligence work as intelligent human or animal mind it makes maximum changes to fulfil the requirements of the concept to achieve the goal. This chapter describes the issues of privacy and security problems in multi-tenancy and maintaining the privacy in multi-tenancy framework using the concept of artificial intelligence.

Keywords: Multi-tenancy, single tenancy, privacy, security, framework, artificial intelligence, complexity, database, tenants, etc.

7.1 Introduction

It is very important to discuss the new challenges coming in the multi-tenancy environment. The extensive use today of the internet ensures that many user tenants share the same data and software for common application. The database is also common to every tenant in a common system,

---

Email: shweta.solanki01212@gmail.com

Neeraj Bhargava, Ritu Bhargava, Pramod Singh Rathore, and Rashmi Agrawal (eds.) Artificial Intelligence and Data Mining Approaches in Security Frameworks, (119–128) © 2021 Scrivener Publishing LLC

but the main problem is to manage all department data privacy and security; with different department tenants it is very difficult to manage (Singh Rathore *et al.*, 2020). So using the artificial intelligence concept we should improve the privacy and security problem.

Using artificial intelligence we must first understand the problem of privacy and security complexity. Then we can analyse the different threads and the complexity in security and quickly identify the solution of the problem (P. S. Rathore *et al.*, 2013).

Using artificial intelligence we find the relationship between the tenant and the common need of resources. Using artificial intelligence concept the privacy and security concept can be increased. As the use of the technology increases, so does the risk of privacy and security (Naveen Kumar *et al.*, 2018). Using artificial intelligence the isolation of each tenant data access area should be maintained. Multi-tenancy is very popular in today's technological market. There are a large number of tenant use common software or the single platform so there is no need to develop many software to many tenants so the cost is also reduced. The multi-tenancy concept is very flexible to maintain, develop and share by many tenants at a time. So the privacy and security concept should be developed fast using artificial intelligence (Neeraj Bhargava *et al.*, 2017).

7.2   Framework of Multi-Tenancy

Multi-tenancy system architecture is that the block of database can be shared with the user or customer or client known as tenant; the tenant uses the block of database for process or for required work that can be done by the tenant.

So many tenants can do their work using the module or the block of code as their required work or task. After completion of the task they can change the task or can change the need of database (Ritu Bhargava *et al.*, 2018).

The main key concept was that the database was not changed; only the part of the database can be used by the many tenants according to their needs. They can update the application in which they work but not change the code or database which is the basic elements of the organization.

All the concept or submodules content features the artificial intelligence concept. Today, many more companies and organizations are using the concept of multi-tenant system structure (Neeraj Bhargava *et al.*, 2018). For growth of the organization and the purpose of computation, which is

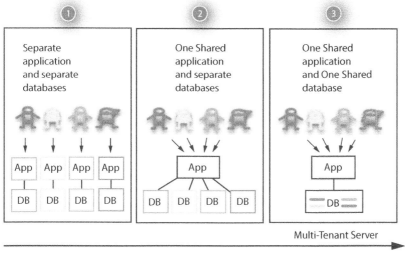

Figure 7.1 Multi-tenancy structure.

the need of today's time for doing fast and accurate work, more development with the correct user or tenant can do the work with the same organization at a different location.

Today's databases have many data, many elements like Table, lists, relation between the table, quires and many more elements in the table. So for maintaining the proper flow of data and the security and isolation of data with many users or tenants the concept of artificial intelligence is required (Pramod Singh Rathore *et al.*, 2017).

Data isolation is the main challenge in the multi-tenant system; the data can be shared by many more users to maintain the work in the proper way but the data should be separated in the many tenants.

The data separation can be done in the basic three conditions given in Figures 7.1 and 7.2. To make implementation with data isolation concept.

Isolate database is very important to use by many users the same data base using the Separate database concept. All tenants can share the many resources, database or block of code for the fulfilment of requirement of organization (N. Bhargava *et al.*, 2017).

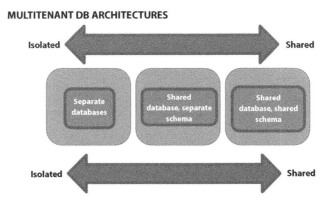

Figure 7.2 Multi-tenancy structure.

7.3   Privacy and Security in Multi-Tenant Base System Using AI

In multi-tenant concept the security and privacy is the big challenge to manage that all data with the concept using artificial intelligence become easier to maintain, and cost should also be reduced using AI algorithms.

All users or organizations demand more security because they all use a common database. The work is different but data is common to security, and privacy is the main requirement of the user. For that purpose they pay money and also demand a high level of security so that all can be done using all the new concept of AI. In artificial intelligence the developer can do much work regarding isolation of data which algorithms use to maintain the isolation of data in the base of user or tenants.

Many times the tenant uses the same table or the same database for the separate work and also saves the work in the common table. So it is very important to isolate the data and maintain the privacy and security between all the tenants. The goal is to work efficiently without any data leak or hacking problem and produce fast and efficient work.

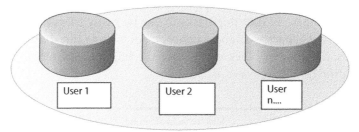

Figure 7.3 Multi-users in system.

As shown in the Figure 7.3, the common database can be shared by many users or companies or multiple tenants. They share the table and common data common schema and common structure. So high security is required to manage the data and ensure that it is not accessible by the wrong tenant.

Using artificial intelligence the different schema are created for each tenant to share the common database but use different schema to make the work more secure and safe.

In artificial intelligence this concept enables many or multiple tenants to create their known separate area in the database; they use the common data but work separately and also store their work in the same database after completion. Other tenants also do the same if the different tenants which are working in the common database module can also do the same; using the AI (artificial intelligence) concept the multiple tenant work will be separated or secure.

So using the artificial intelligence concept the isolation of data in a secure way and also separate schema can create for each tenant the ability to work efficiently using AI algorithms shown in Figure 7.4.

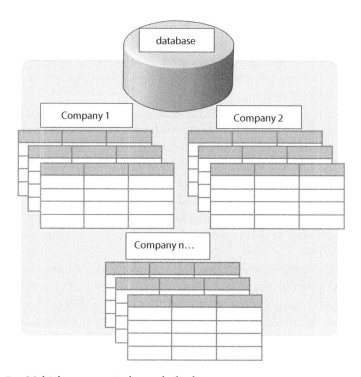

Figure 7.4 Multiple company in the single database.

Continuing with the three concepts of the security of the database in the multi-tenant system using artificial intelligence, the first one is to separate the data and also separate the tenant. In the second, the common data are shared but the structure and schema are separated. But in the third structure and the schema both are common and using the artificial intelligence concept the security and privacy policies are maintained.

So according the schema the AI algorithms depend and using those algorithms the cost depends, according to the algorithm use in the multi-tenant system. There is a high risk of data hacking by the other tenant so Artificial intelligence concept can make all the structure and the schema more secure and the policing of privacy is also stronger and more able to satisfy the tenant.

As we all know, data is very important to all organizations. According the security and safe access the multi-tenant system adopt by the organization.

All control is not manageable by the organization and also not manageable by the tenant itself so for that, artificial intelligence is required to maintain the security and all data manageable schema is prepared by the artificial intelligence concept to make the data base more secure and more manageable (Kumar Sahwal *et al.*, 2018).

Also maintain the cost of the security and privacy policies, because we know security is the big challenge in a multi-tenancy system. Artificial intelligence can meet the big demand of privacy and security concept in the multi-tenancy system.

So a secure environment is demanded in a good multi-tenancy system. And the secure multi-tenant system is today's need. A secure environment is needed to share the structure of database. In database the relationship between the table and the shared table, shared schema, shared data should be isolated between multiple tenants. And the work done by the tenant should be secure and safe submit and update the database by many tenant parallel all work should be done in a safe and secure environment in multi-tenant base system which is provided by the Artificial intelligence concept base algorithms.

Artificial intelligence develops the real-word application for security peruses. AI provides the robust, affordable computing schema to make the multi-tenant structure more secure and the privacy of the tenant should maintained. AI provides good service to the tenant maintenance level of the good security service and prioritizing support in the schema developed by the AI.

In the future millions of development schema for security and privacy policies will be developed in artificial intelligence concept using artificial intelligence algorithms.

7.4   Related Work

In this chapter the work related to the concept of privacy policies in multi-tenancy framework using artificial intelligence concept was examined. The concept of artificial intelligence is used to develop the multi-tenancy system and make it secure and safer, isolated, with privacy and durability for the future.

With the use of artificial intelligence, database hacking and the transition fraud can be found very fast. The cost of maintenance and the security and privacy policy will be reduced using the artificial intelligence concept. There has been much work done in many chapters on the basis of privacy and security of multi-tenancy system using artificial intelligence concept.

This chapter data is useful to find the functional and non-functional parameter of multi-tenant system with respect to artificial intelligence security algorithms.

In this chapter we have discussed in detail the concept of privacy, security, data isolation, durability and the maintenance of the multi-tenancy system structure. We have provided details about the privacy and security concept of the multi-tenancy system framework, and the various concepts of multi-tenancy system use according to the requirements and maintenance of the security and the privacy concept using artificial intelligence.

With the use of artificial intelligence concept we can maintain the security, performance, cost, flexibility factor of multi-tenancy system. This chapter is a contribution to the security and privacy in multi-tenancy system's use of artificial intelligence system. This chapter is usesd to find the maximum solution to make the security and privacy policies in artificial intelligence in the multi-tenancy system more secure.

Understanding the system using artificial intelligence, this literature is used to understand and find the requirement of resources and services and the privacy methods to develop various services like response time, network load, and throughput management services.

7.5   Conclusion

In multi-tenancy system the security and privacy concept using artificial intelligence algorithms are very useful and make the database more secure and allow for safe work. Using AI algorithms or AI concept provides reusability and secure structure to each multi-tenant system. So the data must be accurate, secure, correct, and durable.

Every multi-tenancy system wants to work in privacy and security to work and maintain durability. So using the AI concept provides a secure environment to perform the work and maintains the flow of work in the system. It provides a good and graceful environment.

If the database is not secure the tenant is no longer able to maintain the database and the structure. In today's environment in which many people work remotely at many different locations the artificial intelligence concept is very important.

For the reimplementation of the multi-tenant system and the necessary security in the multi-tenant system it is very important to use the correct artificial intelligence concept to provide the security and privacy to the main database in the multi-tenant framework.

So the artificial intelligence concept use to develop the multi-tenant system to ensure that performance, security, privacy is not lost and durability is maintained.

The multi-tenancy concept is not an easy process to maintain the data isolation, privacy, security, and many other concepts. For using the artificial intelligence concept the security and privacy concept must be developed.

With the help of Artificial intelligence, a multi-tenant system provides secure and developed privacy policies, whch makes working in different databases more stable. The different tenants work in the same database in the isolated data of the same organisation.

References

Singh Rathore, P., Kumar, A., & Gracia-Diaz, V. (2020). A Holistic Methodology for Improved RFID Network Lifetime by Advanced Cluster Head Selection using Dragonfly Algorithm. *International Journal of Interactive Multimedia and Artificial Intelligence*, 6 (Regular Issue), 8. http://doi.org/10.9781/ijimai.2020.05.003

W. Deng, J. Hu, and J. Guo. Extended src: Undersampled face recognition via intraclass variant dictionary. *IEEE Trans. Pattern Anal. Machine Intell.*, 34(9):1864–1870, 5012.

P. S. Rathore, A. Chaudhary and B. Singh, "Route planning via facilities in time dependent network," *2013 IEEE Conference on Information & Communication Technologies*, Thuckalay, Tamil Nadu, India, 2013, pp. 652–655. doi: 10.1109/CICT.2013.6558175

Naveen Kumar, Prakarti Triwedi, Pramod Singh Rathore, "An Adaptive Approach for image adaptive watermarking using Elliptical curve cryptography (ECC)", *First International Conference on Information Technology and*

Knowledge Management pp. 89–92, ISSN 2300-5963 ACSIS, Vol. 14 DOI: 10.15439/2018KM19

Neeraj Bhargava, Abhishek Kumar, Pramod Singh, Manju Payal, "An Adaptive Analysis of Different Methodology for Face Recognition Algorithm", *International Journal on Future Revolution in Computer Science & Communication Engineering (IJFRSCE)*, December 17, Volume 3, Issue 12, pp. 209–212.

Ritu Bhargava, Pramod Singh Rathore, Rameshwar Sangwa, "An Contemplated Approach for Criminality Data using Mining Algorithm", *International Journal on Future Revolution in Computer Science & Communication Engineering (IJFRSCE)*, February 18, Volume 4, Issue 2, pp. 236–240.

Xiaogang Wang and Xiaoou Tang, "Dual-space linear discriminant analysis for face recognition," *Proceedings of the 5004 IEEE Computer Society Conference on Computer Vision and Pattern Recognition, 5004. CVPR 5004*, Washington, DC, USA, 5004, pp. II–II.

Neeraj Bhargava, Pramod Singh, Abhishek Kumar, Taruna Sharma, Priya Meena, December 17 Volume 3 Issue 12, "An Adaptive Approach for Eigenfaces-based Facial Recognition", *International Journal on Future Revolution in Computer Science & Communication Engineering (IJFRSCE)*, pp: 213–216.

www.accenture.com/us-en/insights/artificial-intelligence-index

Slideshare.net. in www.slideshare.net/saa-s-multitenant -database-architecture

Multitenancy in SaaS-PaaS, www.multitenancy-in-saas-paas.wikipaces.asu.edu/

Apprenda PaaS software, www.apprenda.com/library/glossary/definition-multitenant/

Microsoft Developer Network, //msdn.microsoft.com/en-us/library/

Pramod Singh Rathore, "An adaptive method for Edge Preserving Denoising, International Conference on Communication and Electronics Systems, Institute of Electrical and Electronics Engineers & PPG Institute of Technology (2017). *Proceedings of the 2nd International Conference on Communication and Electronics Systems (ICCES 2017)*: 19–20 October 2017.

N. Bhargava, A. Kumar Sharma, A. Kumar and P. S. Rathoe, "An adaptive method for edge preserving denoising," *2017 2nd International Conference on Communication and Electronics Systems (ICCES)*, Coimbatore, 2017, pp. 600–604, doi: 10.1109/CESYS.2017.8321149.

IBM Knowledge Center, www.01.ibm.com/support/knowledgecenter/ssepgg9.7.0/

W. Deng, J. Hu, and J. Guo. Face recognition via collaborative representation: Its discriminant nature and superposed representation. *IEEE Trans. Pattern Anal. Mach. Intell.*, (99):1–1, 5018.

W. Deng, J. Hu, J. Guo, H. Zhang, and C. Zhang. Comments on "globally maximizing, locally minimizing: Unsupervised discriminant projection with applications to face and palm biometrics". *IEEE Trans. Pattern Anal. Mach. Intell.*, 30(8):1503–1504, 5008.

Xiaogang Wang and Xiaoou Tang, "Dual-space linear discriminant analysis for face recognition," *Proceedings of the 5004 IEEE Computer Society Conference on Computer Vision and Pattern Recognition, 5004. CVPR 5004.* Washington, DC, USA, 5004, pp. II–II.

N. Bhargava, S. Dayma, A. Kumar and P. Singh, "An approach for classification using simple CART algorithm in WEKA," *2017 11th International Conference on Intelligent Systems and Control (ISCO)*, Coimbatore, 2017, pp. 212–216. doi: 10.1109/ISCO. 2017. 7855983

Y. Deng, Q. Dai and Z. Zhang, "Graph Laplace for Occluded Face Completion and Recognition," in *IEEE Transactions on Image Processing*, vol. 50, no. 8, pp. 2329–2338, Aug. 5011.

Kumar Sahwal, ..., Kishore, ..., Singh Rathore, ..., & Moy Chatterjee, ... (2018). An Advance Approach of Looping Technique for Image Encryption Using In Commuted Concept of Ecc. *International Journal of Recent Advances in Signal & Image Processing*, 2(1).

Yuille, A.L., Hallinan, P.W. & Cohen, D.S. Feature extraction from faces using deformable templates. *Int J Comput Vision* 8, 99–111 (1992). https://doi. org/10.1007/BF00127169.

Z. Lei, C. Wang, Q. Wang and Y. Huang, "Real-Time Face Detection and Recognition for Video Surveillance Applications," *2009 WRI World Congress on Computer Science and Information Engineering, Los Angeles, CA, 2009*, pp. 168–172.

8

Biometric Facial Detection and Recognition Based on ILPB and SVM

Shubhi Srivastava*, Ankit Kumar and Shiv Prakash

Centre for Advanced Studies, Dr. A.P.J Abdul Kalam Technical University, Luck-now, Uttar Pradesh, India

Abstract

Biometric security has long been a trending zone that satisfies the need for a significant level of security and control. Among all the existing technologies, face detection is one of the most utilized and adjusted innovations. The identification failure of a user's identity is a big concern. In this chapter, a novel approach for biometric recognition has been introduced in which the application of ILBP (Improved Local Binary Pattern) for facial feature detection is discussed which generates improved features for the facial pattern. It allows only an authenticated user to access a system, which is better than previous algorithms. Previous research for face detection shows many demerits in terms of false acceptance and rejection rates. In this paper, the extraction of Facial features is done from static and dynamic frames using the Haar cascade algorithm. Then, the ILBP method which works on local pixel values of an image is applied for feature extraction, and finally, the SVM (support vector machine) is used for classification of those features. The objective of this paper is to provide the best recognition results from images that are taken randomly and may possess noise. This paper achieved an accuracy of 97.90% for correct recognition and with less time complexity. It can be used in crime investigation, security cameras, digital forensics, etc.

Keywords: Face detection, ILBP, SVM, feature extraction, haar cascade algorithm

8.1 Introduction

The system authentication-based facial biometric features are most commonly used in many real-time IoT devices. Facing many threats related to

---

**Corresponding author*: shubhisrivastava1095@gmail.com

Neeraj Bhargava, Ritu Bhargava, Pramod Singh Rathore, and Rashmi Agrawal (eds.) *Artificial Intelligence and Data Mining Approaches in Security Frameworks*, (129–154) © 2021 Scrivener Publishing LLC

document fraud, identity threat, cybercrime, terrorism and many more, new technologies have been implemented. Among them is biometrics, which is used in authenticating and identifying an individual using his biometric characteristics such as fingerprints, face recognition, iris detection, etc. Features of a biometric subject are very crucial for its correct recognition. Biometric identity is the user's unique and permanent identity and its recognition failure depends on many circumstances such as dataset, algorithm, feature learning, pirated features, etc. The correct and consistent recognition of the biometric subject is a challenging task. All the existing research on biometric recognition requires some unique key-points (A. S. Georghiades *et al.*, 2001) that sufficiently define the proper authentication of an individual. Biometric data like fingerprint, iris scan, palm print, and face subject are commonly used for authentication research and are widely trusted since they contain unique features. A variety of biometric images may contain noise and the system may fail to extracted proper key-point features from a low-quality biometric image which causes its recognition failure (B. Heisele, P. Ho and T. Poggio, 2001). This work endorses the concept of biometric authentication (C. Liu and H. Wechsler, 2000) using biometric facial features. This work not only provides the security of authentication but also satisfies the consistency in recognition under a variety of dataset.

The failure of recognition of such features seems to be under two wide cases. First, the false acceptance rate in which an unauthorized user gets access and second, a false rejection rate in which an authorized user may fail to gain access. Such apprehension in the security system is mostly contemplated as recognition error caused by inappropriate image detection, low-quality image analysis, low-quality pixel information, etc. The face recognition system is viewed under various classifier algorithms such as KNN, AdaBoost, Haar cascading, etc. All such previous defined models work well for a particular dataset only. The diversity in facial images may not easily be analyzed by the model and end up with an error. Such diversity in the dataset may contain low-quality pixels image, noised image, filtered image, and other unwanted random variations. To make dominance on such variety, the proposed technique uses the concept of improved local binary pattern (ILBP) histogram method in which all the undesired noise on facial capture is extirpated with the help of features reinforcement or suppression based on binary intensity values. A robust technique for classification of improved facial features is rationalized by using support vector machine algorithm (SVM) in which the reinforced facial features of an image are analyzed based on their geometric values and the hyper plane is decided by SVM to classify the discriminate features. The decision of correct belongingness of the test features is taken by SVM which is based on training features. This rigid combination of technique is undoubtedly convenient and less complex for recognition of undesired variety in facial images. The proposed model also successfully recognizes the multiple facial subjects distinctively, which is captured in a single frame.

8.1.1   Biometric

Biometric is the science of using an individual's physical and behavioral characteristics in order to authenticate and identify a person. Biometric authentication is defined as comparing data saved in the database as biometric template to that of an individual's characteristics to find the resemblance (Frank Y. Shih, Chao-Fa Chuang, 2008).

- Database is prepared by keeping a record of biometric template of registered individuals.
- Then data saved is compared with the individual's biometric data for his identity authentication.
- Biometric identification is the process of knowing an individual's identity.
- Here a person's biometric data is taken which can be any of his biometric characteristics.
- This data which is stored is then compared to the other biometric data of persons stored in the database. If any of the data matches, the user is identified; otherwise a person is not an authenticated user.

8.1.2   Categories of Biometric

There are two categories of biometric.

Physiological measurements
Physiological measurements can be either biological or morphological. It consists of morphological shape of a face, fingerprint, palm print, the eye (retina and iris), vein pattern, etc. Biological analysis can be done by saliva, DNA, urine, by a medical team and a police forensic.

Behavioral measurements
Common behavioral measurements that can be used are voice recognition, body gestures, gait, keystroke dynamics, etc. Different techniques used for these measurements are ongoing research topics.

Physiological measurements usually offer the benefit of remaining more stable throughout the life of an individual.

This work is conducted on biometric facial features in order to verify a person's identity. Face recognition is one of the most used techniques in image analysis. In this, facial features of a person are extracted and then correlated with the images stored in the database of the system. Face is detected from an image and video. After its detection from an image features are positioned and then a face print is taken from facial features. Finally, using the object classification method, data extracted is compared with the data saved in the database and the face recognition method is completed.

8.1.2.1 Advantages of Biometric

- It is universal as it can be found in any individual.
- It is unique as every individual has its own characteristics.
- It is a permanent characteristic.
- It can be recorded in any database.
- Measureable data.
- Forgery proof unlike signature of an individual.

8.1.3 Significance and Scope

To allow only an authenticated user to access confidential data, biometric technology finds its way into many electronic devices, BFSI, criminal identification, banks, home security, defense, etc. It is even being used by many public and private offices in the form of devices such as fingerprint, voice recognition, face recognition, etc. Significance of biometric is increasing day by day as the need of security is an important concern. Taking an example of Maharashtra Government, which has created a database of criminals with their figureprint, retina scan, and face recognition, which helps police in apprehending them without delay in the investigation (G.R.S Murthy, R.S. Jadon, 2009). There are many such examples of use of biometric by government. One more such example is Adhar cards, which save an individual's data such as figureprint, retina scan, etc., in a database.

8.1.4 Biometric Face Recognition

The model of face recognition is not easy because of its complexity and multidimensionality. Computation constraint is always there in face recognition model. Face recognition is as similar as any pattern recognition model but face recognition deals with the biometric features. The focus of the model is always over detailed features of facial subject in order to distinguish one face from another. The aim of face recognition is the identification of unique feature vectors that provide accuracy. Handling high visual facial information is quite achallenging task in face recognition. This high visual information can be obtained from live video recording in which high-quality images are captured which takes more computation time to execute the recognition. The face recognition takes the unique set of characteristics which resides in the eigen values of the subject. These values are trained in the model and the features are classified in unique classes. These eigen feature are considered to be a series of stable values which are ineffective towards any modification. And so these values are helpful for face recognition.

The modification can be introduced in the facial subject in terms of any image processing attacks or by any illumination effect. But these eigen values may be able to tolerate such variation and do not alter. Hence, these values are easily used in face recognition. The recognition system involves the correct localization of eyes, nose, mouth, face outline, etc. These subjects are helpful to connect the entire facial pattern recognition by the model. These traits are appropriate to establish the relationship among the different parameters of the facial pattern in order to achieve correct recognition. Earlier published technology uses multiple automated and semi-automated recognition strategies based on normalized distances metric among the feature points. The difficulties of multiple views are also addressed by such earlier techniques. The objective of recognition is accomplished by finding adequate relationship among the various traits and subjects of facial part. The feature training is one of the crucial tasks in which features are learned by the entire network of the model. Kohonen *et al.* discussed a training system which learns non-linear units of features and also uses back-propagation algorithm to rectify any error in feature learning. The face identification along with expression is also discussed by Stonham *et al.* Another research on face recognition uses multi-resolution of template matching which uses smart sensing algorithm and it was given by Burt *et al.* This kind of recognition satisfies real-time identification and processing. Face recognition is based on matching the test features with the trained features and then recognition of the test features gets established in the respective classes by the classifier. Graph matching is one of the efficient approaches in which structures of dynamic link are created based on featured mapping. Such matching can be seen in ANN classifier which network is formed between the nodes.

The feature matching is also done using geometrical distance metric in which distance between the two pixel values are calculated using Euclidean distance and the matching is said to be successful only if the distance is small otherwise rejected. The matching is also done using intensity of the pixel values of the features of an image. The two images are said to be equal if their cross-ponding feature pixel intensities are closer to each other. The mapping criteria are different for each model.

Face recognition is used in authentication, criminal investigation, film processing, image forensic, etc. So it is our objective to develop a model which is able to identify each face uniquely in crowd and in any variation. The eigen features–based approach is very commonly adopted by researchers like Pentland *et al.* These eigen values are helpful in constructing the facial subject. It does not consist any unwanted variations and effect caused during bad illumination or by non-symmetrical positions of person's face. The visual of facial subject is matched in many ways. One such way is probabilistic-based matching in which the intensity of pixel values is mapped using Bayesian algorithm. This algorithm is used to predict the matching based on a predefined mapped values. It also deals with the variation on images caused by several kinds of noise. This type of algorithm uses a probability density function for every individual class and also uses an optimal version of eigen values which is very effective in matching. Further, face recognition is basically carried out by using local binary pattern or LBP features which contain the texture information of the image. These LBP features of a facial image are formulated with the help of neighboring pixels values of the center pixel.

The LBP features are easily plotted by histogram representation which describes the texture information of the image. Face recognition scenario contains majorly three things, named as face detection, normalization and face identification. The first part, face detection in facial portion, is detected from the entire image and segmented out. The image of a person contains images of background, hair, clothes. etc. The interest portion is face only. So the detection of facial portion from the entire image is very crucial and it also works in face detection in crowd. The detection is done by several algorithms like Haar cascade, Viola-Jones, etc. After the detection process, the normalization of image is carried out in which the pixel of the image is normalized so that we can avoid any unwanted variations and noise. The last step toward face recognition is recognition itself in which normalized features are trained and classified in several classes and hence the test features are recognition-based on the described classes by the classifier.

The basic flow of facial pattern recognition is also described by the flow chart described in Figure 8.1 which helps us to understand the working.

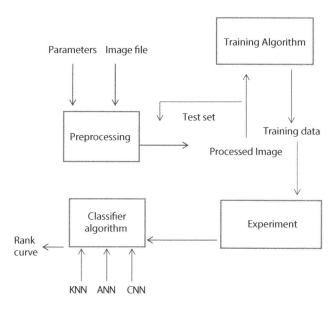

Figure 8.1 Flow chart of general face recognition process.

With the above diagram, it is clearly seen how the entire face recognition process will work. First, the captured image file and the relevant parameters like its size and resolution is to be passed in the preprocessing unit. The function of preprocessing is to remove any unwanted noise and remove disparity between the images. Basically, it maintains the symmetry of the facial image. Also, it does the segmentation if necessary. The captured facial images may have additional unwanted information like background image or any other side image. So, the facial portion is requiring to be segmented out from the entire supplied image. Then the images are divided into test and train set. The training set contains large proportion as compared to test set. The training block is used to learn the features of the preprocessed trained images. These training features are then passed in the experiment block. So, the experiment block is used to evaluate the features of training features and pass these features to the classifier algorithm. The classifier algorithm uses any type of classifier like KNN, ANN, CNN, etc., which are used to classify each trained features into their unique classes (P. S. Rathore *et al.*, 2013). The preprocessed test features are then passed to the classifier algorithm which is used to predict the belonging classes of the test subjects. The classifier job is not only to classify the trained features but also predict the recognition of the test cases in the given classes. The correct and accurate recognition results are formulated in term of rank curve which is nothing but the ROC curve which shows the number of test features matched correctly in their respective classes.

8.1.5   Related Work

Face recognition techniques are widely explored earlier based on various machine learning or deep learning approach in which the analysis of features vector takes place. Such a technique named as support vector machine (Singh Rathore *et al.*, 2020) (G.R.S. Murthy, R.S. Jadon, 2009), linear discriminative analysis, Laplacian algorithm, evolution pursuit, etc., that have accessed face recognition successfully. All such techniques are a little complex and more time consuming for vague or high-quality images. And they do not guarantee the recognition accuracy for the diversified variations in images. In addition, Hallinan *et al.* proposed a face recognition technique based on eigenface features that were undertaken with the variations of lighting conditions, but this method could not guarantee the recognition blur images that contain the least facial information. Belhumeur *et al.* also discussed face recognition based on the 3D linear subspace technique which works well for variable lighting and orientation. But this technique is a little complex and has high computation time as it works on a 3D subspace. The Model-based on SPD and Grassmann manifolds consider all the images with equal quality. These methods are good for the low-quality images as it enhances all the images to an equal level. But still, it does not justify the recognition for a frame captured with multiple faces. Satisfying the recognition of facial data distinctively in-crowd is one another major issue in recognition systems. (Caicedo *et al.*, 2015) also discussed an effective method for face localization using reinforcement learning, but it is a little complex as it works on the basis of decision tree and hence it consumes more time to localize the face subject. The proposed technique of this paper uses Haar cascading classifier to localize face portion from the captured image and this technique is much simpler as it uses Haar features for masking the rectangular block of face subject (N. Bhargava *et al.*, 2017). Face recognition relied on manifold and sparse or linear subspace are also in some of the obsolete techniques which fail to tolerate false rejection and false acceptance error rates.

8.1.6   Main Contribution

The contribution of the paper lies not only with the static images but also with the dynamic real-time video frames. Data from the videos can be captured, detected and classified with a minimum error rate. Facial images are extracted using the Haar cascade algorithm in which it locates the haar features on the facial subject of the entire image. The proposed technique introduces a new combination of the robust design of ILBP and SVM. The ILBP algorithm provides an improved version of local features of facial patterns to the SVM model. The SVM is used as a classifier that provides the best results in terms of recognition of such improved local features with less time complexity. Face detection, feature extraction, and recognition is proficiently done using advanced algorithms that maintain the recognition accuracy over the variety of noise in images while preserving the time complexity. The accuracy of the recognition is found to be 97.90% which is better than the other existing algorithm (J. C. Caicedo *et al.*, 2015).

8.1.7   Novelty Discussion

Improved versions of local binary features are extracted from the facial images of users. These features are very stable and do not cause any distortion on any modification. The features are not easily traceable. The concept behind this using these features is their consistency. These are refined features and help to improve the authentication perseverance of the model for random images also. The idea of using the ILBP pattern of facial features comes with the limitation in the classification model. In the earlier models, the recognition and the classification of facial features seem to fail for random facial images which may contain random noise. In simple words, the facial images that are captured randomly in real time from a crowd may contain many unwanted variations like lighting effect, background illumination, blurring, improper orientation, improper poses, angels, etc. In other simple words, the randomly clicked images of users that are clicked in real time from a crowd are different from a regular clicked image that is clicked in the symmetrical background with constant lighting and distance. Therefore, the challenging task is to secure recognition of such facial images that are clicked randomly without any symmetrical parameters and hence contains random variations (Bharat Singh *et al.*, 2013). The proposed work contributes to securing the recognition and classification of such images. Therefore, the proposed model is using ILBP features that are extracted from such random facial images, and by using SVM, the classification work takes place. The idea behind using SVM is to save computation time because such facial images are not previously defined. The prediction of their processing time is not always the same as it depends on the variation that an image is loaded with while taken with a camera. SVM is a lightweight program and able to generate hyper-plane easily for a large number of feature vectors (J. Zhang, L. Wang *et al.*, 2016).

Although previous techniques are capable of producing a recognition accuracy rate that is good enough, the proposed model gives a 97.90%

accuracy rate of recognition which is an acceptable criterion and also competing with some of the latest researches of 2018, 2019, and 2020 in the face recognition model. The novelty of the proposed model is not limited to the accuracy rate but it extends up to the dataset also. The proposed work uses a self-made dataset that contains random clicked images of a different person, as discussed earlier. These images are not clicked with a specific distance or in specific environmental conditions. Therefore, the proposed model secures the consistency of recognition of such images which contains random variations (like in real-time footage) with an accuracy of 97.90%. The main difference between other techniques and our proposed model is not only the accuracy rate but also the type of dataset and features taken. In other existing techniques, researches were in a symmetrical dataset containing symmetrical images that are clicked in specific environmental conditions. The consistency of recognition by the existing model may not be robust enough to process those images containing large random variations. The recognition stability of all machine learning models is based on their learning features or training. Those existing models can work only if necessary training is provided to them. But, this is such an overhead to train a machine learning model again and again for variable datasets. To get rid of this, the proposed model uses ILBP features that are most stable and refined features which are extracted for non-symmetrical images, and they are not expected to modify under the noise. Another novelty of the proposed work is to detect the facial portion from the clicked images. The clicked images contained a person's other body parts and background objects also. As the proposed work uses non-symmetrical random clicked images, an image of a person may carry other side objects or other people around. Any random clicked image from a street in real time contains objects, other people, vehicles, dark background, lighting effect, etc. Therefore, another challenging task is to detect and segment out the required facial portion of a person from the entire clicked image. The proposed model uses the Haar-cascade classifier to detect the facial portion from an image distinctively. The technique first plots the Haar features of the facial portion and by detecting the shape of nose and eyes, it marks a rectangular boundary around the facial portion. Therefore, in the case of a crowd, each facial detail can easily segment out without any mismatch error. In existing work, the detection task is not challenging as the images are taken in specific environmental conditions (Jyh-Yeong Chang *et al.*) (Pramod Singh Rathore *et al.*, 2017).

The remaining paper is divided into the following: Section 2 defines the proposed methodology, section 3 contains experimental data, section 4 deals with the conclusion, which is followed by the references.

8.2   The Proposed Methodolgy

8.2.1   Face Detection Using Haar Algorithm

The detection of a facial subject from the entire image is done by the Haar cascade algorithm. The detection of a facial subject depends on the features of the image that this algorithm uses. This classifier was given by Viola and Michael Jones (K. A. Gallivan *et al.*, 2003). The Feature values are calculated by a Haar classifier using a rectangle integral in which it multiplies the weight of the each rectangle with its area value and sum-up all of them. The algorithm works on detecting some essential features of a face and draws a rectangle around it. The algorithm is trained with some positive and negative images and then depending on those images features are detected (Naveen Kumar *et al.*, 2018). Figures 8.2 and 8.3 show the detected and segmented facial subject from an input image.

Haar features are extracted by surrounding required pixels with a rectangular structure. Rectangular structure covers up all the required pixels

Figure 8.2 Detecting facial feature from an image.

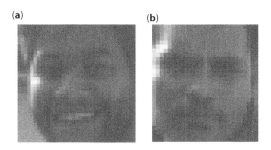

Figure 8.3 Extracting detected images.

at four corners and sums up all of them (Neeta Sarode and Shalini Bhatia, 2010). Computation process can be seen from the equation given below.

$$\sum_{a=j0}^{j1} W \sum_{b=k0}^{k1} I(a,b) = F(j1,k1) - F(j0,k1) - F(j0,k1) + F(j0,k0)$$

Where I is an integral image and sum of intensities are ranging from (x0, y0) to (x1, y1) which can be seen from Figure 8.4.

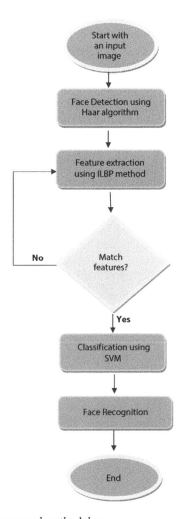

Figure 8.4 Flowchart of proposed methodology.

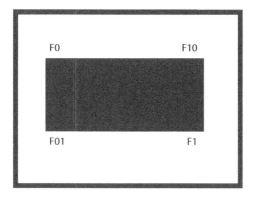

Figure 8.5 Rectangular integral.

All the four values of rectangle are summed up in order to get the particular Haar features shown in Figure 8.5.

8.2.2 Feature Extraction Using ILBP

This paper proposes a new method to extract the image's confine features. It is not desirable to obtain a high-dimensional matrix of an image. Only the crucial confine features of an object are necessary to extract. The confine characteristics are obtained by using ILBP having low dimensionality (W. Deng *et al.*, 2012). The selected images for the experiment may suffer from diverse variations against the illumination effect and some certain noises are the result of size, translation, rotation, and random expression. These factors may restrict in recognition of the confine features of an image. The proposed model introduces the ILBP algorithm which extracts the local definitions of facial key-points under the variable disturbance. The technique of Improved Local Binary Patterns is rooted in 2D pattern determination. In ILBP average value of pixels is compared with nearby or neighborhood pixel values in order to get the least difference with all the other pixel values. This helps in getting the distance of average pixel value with all other pixel values which returns a better pattern. In ILBP whole 3 x 3 neighborhood pixels are thresholded by an average grey scale value and it provides 2^9-1 possible patterns. The work of the process is defined equation 2 (Xiaogang Wang *et al.*, 2004).

$$(x) = \begin{cases} 1, & x \geq 0 \\ 0, & x < 0 \end{cases}$$

$$f_{ILBP}(x) = 2^8 \xi(Ic - S) + \sum_{j=0}^{7} 2j\, \xi(Ij - S) - 1$$

where Ic is the center pixel value Ij is a neighborhood pixel value and S is the average greyscale value which is calculated in equation 3 (Y. Deng *et al.*, 2011).

$$S = 1/9\left(Ic + \sum_{j=0}^{7} Ij \right)$$

ILBP represents the feature vector in the form of a binary pattern of 0 and 1. This binary pattern is converted into decimal form and is stored in a matrix representing the feature vector, as shown below.

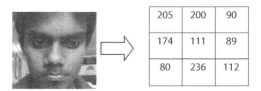

Gray Scale Image   **Pixel value of Gray Scale**

(205+200+90+89+112+236+80+174+111)/9=149.67

(111-149.67) = negative value = 0

(112-149.67) = negative value = 0

(205-149.67) = positive value = 1

(174-149.67) = positive value = 1

(89-149.67) = negative value = 0

(80-149.67) = negative value = 0

(90-149.67) = negative value = 0

(236-149.67) = positive value = 1

(200-149.67) = positive value = 1

$(010100011)_2 =$ (163_{10}) ⟶ REPRESENTS PATTERN

From the extracted facial images, the respective feature vectors are extracted in the form of binary pattern and after converting it to decimal pattern it is stored in matrix. The features are represented by histogram as given in Figure 8.6.

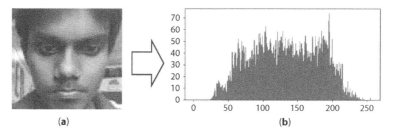

(a)   **(b)**

Figure 8.6 (a) Segmentated image (b) Feature histogram generated by ILBP.

8.2.3   Dataset

The proposed technique is tested on our own dataset which we created using random clicks. These images do not have any symmetry and contain noise in it. The database consists of 200 images. We are having 10 classes each having 20 images. To examine the robustness of our technique we also consider images from video frames. All the images are saved in a single directory. Database images are converted to grayscale (Yuille, A.L. *et al.*, 1992) and then the feature extraction process is carried out. Some images from the dataset are enrolled in the model for training purpose to learn the features and then other random images are used for testing purposes. If the image provided is found from the trained images then the face will be recognized; otherwise it will output it as an unknown person. Database sample images are shown in Figure 8.7.

8.2.4   Classification Using SVM

In this paper, we used SVM to train our model, which provided better accuracy. We considered 200 images which are divided as a total of 10 classes each having 20 images. These images are taken from random clicks and some of them are captured from video frames also. Images taken have different variations, expressions, illumination and also have noise caused by the external factors. After feature extraction, classification is being performed by the SVM (support vector machine). SVM basically forms

Figure 8. 7 People's dataset images.

a hyper-plane that separates the discriminant features. Assume, (X_i, Y_i) is the set of feature sample in which $Y = \binom{+1}{-1}$ is a class label. The equation of hyper-plane is given below:

$$(w \cdot X) + b = 0$$

The optimal classification function of SVM algorithm is given below.

$$g(x) = sng\left(\sum_{i}^{n} Y_i \, a_i^0 \, (X_i \cdot X) - b_0\right)$$

Here, X_i is the support vector or SVs, a_i^0 is the corresponding language coefficient, b_0 is a threshold value. This equation establishes the linearly separable cases for each discriminant features.

In SVM classification, the high dimension space is mapped with the input vectors with the help of non-linear transformation. In this way, an optimal hyper-plane can be obtained. If an inner product function or a kernel function is used instead of using dot product in optimal classification then the corresponding discriminate function will result in the following:

$$g(x) = sng\left(\sum_{i}^{n} Y_i \, a_i^0 \, K(X_i \cdot X) - b\right)$$

It shows a discriminate function of the Support Vector Machine. It is possible to generate various nonlinear discriminate SVM functions using different kernel function $K\,(X_i, X)$. The following generated functions are given below:

Polynomial function:

$$K\,(X_i, X) = [(X \cdot X_i) + 1]^d$$

Here, the SVM is a classifier of polynomial of degree d.
Radial function:

$$K\,(X_i, X) = \exp\{-|X - X_i|^2/\alpha^2\}$$

Here, SVM act as a Gaussian RBF classifier.
Sigmoid function:

$$K(X_{i,} X) = \tanh(v(X_{i}.X)+c)$$

Here, SVM act as a multi-layer perceptron.
$X_i = (x_1, x_2, x_3 ....x_m)$ are the support vectors. The K $(X_{i,} X)$ indicates the kernel function. The result of the SVM is:

$$Y = sgn\left(\sum_{i=1}^{n} Y_i a_i K(X_{i,} X) - b\right)$$

Using distance metric similarity between the tested features and trained features are calculated and finds the Euclidean distance between the feature values and then it is compared to threshold value. Accuracy is increased by changing threshold value. Distance calculated should be less than the threshold value in order to show the similarity. Therefore, on the basis of similarity between the individual classes images are classified to their respective class. Each class separately belongs to an individual identity which signifies the true belongingness.

8.3   Experimental Results

We have created our own dataset of 200 images with 10 individual classes each having 20 images of different peoples. The dataset also contains the images captured from video frames. Each image is then annotated with different labels. The dataset is divided in the ratio of 60:40 for training and testing sets. 120 images are used for training of the SVM model and 80 images are used for testing purposes. Beginning with the collection of dataset and extraction of features using ILPB feature vectors are calculated and then finally faces are detected and classified into their respective classes. Then tested images are given as an input and if their features are matched with the features of trained images then it is classified as a known image and the belonging class; otherwise, the model recognizes it as an unknown face as shown below.

Figure 8.8 Unknown face.

Figure 8.8 shows an image which is not present in dataset and is tested as unknown person. The experimental results are divided into three following sections:

- Detection of facial subject
- Extraction of Features
- Recognition of Face

8.3.1   Face Detection

Faces from the image are detected and extracted using the Haar cascade classifier. A sample of the detected and segmented images is given in Figure 8.9 and Figure 8.10, respectively.

8.3.2   Feature Extraction

In this part features are extracted from facial part of the images using ILBP method, which is stored in the form of feature vector matrix.

Table 8.1 shows the samples of detected and extracted images and their respective ILBP features that are represented by histogram.

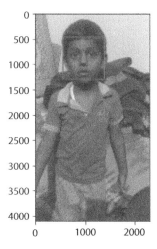

Figure 8.9 Face detection.

Figure 8.10 Segmented facial image.

Figure 8.11 shows a comparison between resulting features that are generated from two different methods, i.e., improved LBP and simple LBP [17]. From Figure 8.10, it is concluded that the ILBP algorithm generates improved features as compared to the features which are extracted by a simple LBP algorithm.

8.3.3   Recognize Face Image

Recognition of facial features and then its classification is the final output of the experiment. If the features match with the trained features in the database then the output will be obtained from the database as the recognized face. The ROC curve given in Figure 8.12 shows the accuracy of feature matching which is done by SVM algorithm.

Table 8.1   Sample of featured extraction using ILBP method.

Samples of images from dataset	Extracted images (RGB)	Feature histogram

(*Continued*)

Table 8.1 Sample of featured extraction using ILBP method. (*Continued*)

Samples of images from dataset	Extracted images (RGB)	Feature histogram

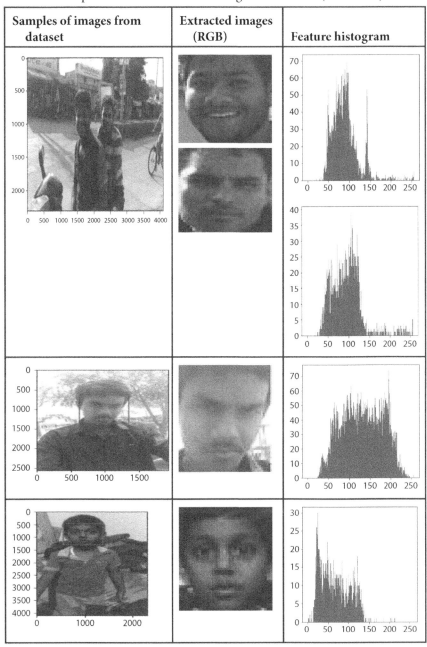

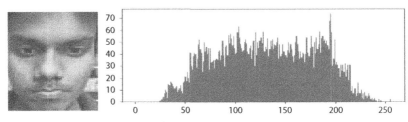

(a) Sample of a segmented image and its feature histogram using ILBP algorithm

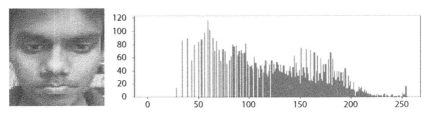

(b) Sample of a segmented image and its feature histogram using the simple LBP algorithm

Figure 8.11 Comparison between feature that are extracted using (a) ILBP method and (b) simple LBP method.

According to Figure 8.12, the accuracy of the model over the test dataset is found to be 97.90%. Some of the tested images consist of unknown data. The error rate [18] in the recognition is 2.10% only. Hence, the proposed model is able to prove good recognition accuracy for different varieties of images that may contain unpredictable noise.

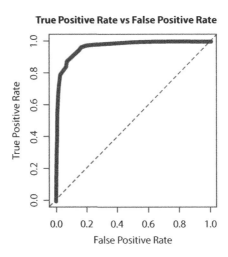

Figure 8.12 ROC curve for face recognition.

Table 8.2 Comparison table.

S. no.	Techniques	Accuracy/Result
1	Rough Contour and Estimation Routine based face recognition	92.1%
2	Principle component analysis based face recognition	83%
3	AdaBoost and advance component analysis based face recognition	95.50%
4	Radial Symmetry Transform	83%
5	SVM based Recognition	95.71%
6	The proposed technique	97.90%

Table 8.2 shows the comparison table in which the proposed technique is compared with other published techniques and shows the good results as compared to others.

8.4 Conclusion

In this paper, the ILBP with SVM is used mainly to identify the facial features of an individual. The experiment is done by applying ILBP along with SVM to classify the results with the best accuracy. So for this, we have tested our approach on multiple images and we came to the conclusion that with ILBP features can be extracted in a better way, which leads to a better accuracy of 97.90%. And also the SVM algorithm is a strong classier which trained the dataset efficiently and classified the tested data accurately and with less error rate. Its accuracy will increase if its training dataset is increased.

One of the most well-known uses of face recognition is for security. Law implementation workforce can utilize this innovation to recognize and distinguish people by examining anybody entering a medical clinic. They would then be able to compare every individual with a rundown of recognized people. The innovation can likewise be utilized in medical clinics to recognize people who may be drug searchers or people recently discharged whom the emergency clinic no longer permits to get to the office. Hospitals can likewise utilize face recognition innovation to distinguish designs that involve generally speaking insights around guests and patients dependent on gender and age. This framework can enable the office to follow patients without utilizing physical GPS beacons. This can prove to be useful to find patients inside a nursing home or in outpatient or assisted living offices.

References

A. S. Georghiades, P. N. Belhumeur and D. J. Kriegman, "From few to many: illumination cone models for face recognition under variable lighting and pose," in IEEE Transactions on Pattern Analysis and Machine Intelligence, vol. 23, no. 6, pp. 643–660, June 2001.

Singh Rathore, P., Kumar, A., & Gracia-Diaz, V. (2020). A Holistic Methodology for Improved RFID Network Lifetime by Advanced Cluster Head Selection using Dragonfly Algorithm. *International Journal of Interactive Multimedia and Artificial Intelligence*, 6 (Regular Issue), 8. http://doi.org/10.9781/ijimai.2020.05.003

P. S. Rathore, A. Chaudhary and B. Singh, "Route planning via facilities in time dependent network," *2013 IEEE Conference on Information & Communication Technologies*, Thuckalay, Tamil Nadu, India, 2013, pp. 652–655. doi: 10.1109/CICT.2013.6558175

N. Bhargava, S. Dayma, A. Kumar and P. Singh, "An approach for classification using simple CART algorithm in WEKA," *2017 11th International Conference on Intelligent Systems and Control (ISCO)*, Coimbatore, 2017, pp. 212–216. doi: 10.1109/ISCO. 2017. 7855983

B. Heisele, P. Ho and T. Poggio, "Face recognition with support vector machines: global versus component-based approach," Proceedings Eighth IEEE International Conference on Computer Vision. ICCV 5001, Vancouver, BC, Canada, 2001, pp. 688–694, vol. 2.

C. Liu and H. Wechsler, "Evolutionary pursuit and its application to face recognition," in IEEE Transactions on Pattern Analysis and Machine Intelligence, vol. 22, no. 6, pp. 570–582, June 2000.

Frank Y. Shih, Chao-Fa Chuang, Patrick S. P. Wang "Performance Comparisons Of Facial Expression Recognition In Jaffe Database" *International Journal of Pattern Recognition and Artificial Intelligence* Vol. 22, No. 3 (2008) 445–459.

G.R.S Murthy R.S.Jadon "Effectiveness of Eigenspaces for Facial Expressions Recognition" *International Journal of Computer Theory and Engineering*, Vol. 1, No. 5, December, 2009, pp. 1793–8501.

J. C. Caicedo and S. Lazebnik. Active object localization with deep reinforcement learning. In *ICCV*, pages 2488–2496, 2015.

J. Zhang, L. Wang, L. Zhou and W. Li, "Learning Discriminative Stein Kernel for SPD Matrices and Its Applications," in IEEE Transactions on Neural Networks and Learning Systems, vol. 27, no. 5, pp. 1050–1033, May 2016.

Jyh-Yeong Chang and Jia-Lin Chen, "Automated Facial Expression Recognition System Using Neural Networks" *Journal of the Chinese Institute of Engineers*, Vol. 24, No. 3, pp. 345–356 (2001).

Bharat Singh, Ravinder Singh and Pramod Singh Rathore. Article: Randomized Virtual Scanning Technique for Road Network. *International Journal of Computer Applications* 77(16):1–4, September 2013.

K. A. Gallivan, A. Srivastava, Xiuwen Liu and P. Van Dooren, "Efficient algorithms for inferences on Grassmann manifolds," IEEE Workshop on Statistical Signal Processing, 5003, St. Louis, MO, USA, 2003, pp. 315–318.

Neeta Sarode, Prof. Shalini Bhatia, "Facial Expression Recognition", *International Journal on Computer Science and Engineering*, Vol. 02, No. 05, 2010, 1552–1557.

Texture Description Through Histograms of Equivalent Patterns Antonio Fernández · Marcos X. Álvarez · Francesco Bianconi Published online: 12 September 2012 © Springer Science+Business Media, LLC 2012.

W. Deng, J. Hu, and J. Guo. Extended src: Undersampled face recognition via intraclass variant dictionary. *IEEE Trans. Pattern Anal. Machine Intell.*, 34(9):1864–1870, 2012.

W. Deng, J. Hu, and J. Guo. Face recognition via collaborative representation: Its discriminant nature and superposed representation. *IEEE Trans. Pattern Anal. Mach. Intell.*, (99):1–1, 2018.

Pramod Singh Rathore, "An adaptive method for Edge Preserving Denoising," International Conference on Communication and Electronics Systems, Institute of Electrical and Electronics Engineers & PPG Institute of Technology (2017). *Proceedings of the 2nd International Conference on Communication and Electronics Systems (ICCES 2017)*: 19-20 October 2017.

W. Deng, J. Hu, J. Guo, H. Zhang, and C. Zhang. Comments on "globally maximizing, locally minimizing: Unsupervised discriminant projection with applications to face and palm biometrics". *IEEE Trans. Pattern Anal. Mach. Intell.*, 30(8):1503–1504, 2008.

Naveen Kumar, Prakarti Triwedi, Pramod Singh Rathore, "An Adaptive Approach for image adaptive watermarking using Elliptical curve cryptography (ECC)", *First International Conference on Information Technology and Knowledge Management* pp. 89–92, ISSN 2300-5963 ACSIS, Vol. 14 DOI: 10.15439/2018KM19

Xiaogang Wang and Xiaoou Tang, "Dual-space linear discriminant analysis for face recognition," Proceedings of the 2004 IEEE Computer Society Conference on Computer Vision and Pattern Recognition, 2004. CVPR 2004., Washington, DC, USA, 2004, pp. II–II.

Prof. Neeraj Bhargava, Pramod Singh, Abhishek Kumar, Taruna Sharma, Priya Meena, "An Adaptive Approach for Eigenfaces-based Facial Recognition", *International Journal on Future Revolution in Computer Science & Communication Engineering (IJFRSCE)*, December 17, 2017, Volume 3 Issue 12, pp. 213–216.

Y. Deng, Q. Dai and Z. Zhang, "Graph Laplace for Occluded Face Completion and Recognition," in IEEE Transactions on Image Processing, vol. 50, no. 8, pp. 2329–2338, Aug. 2011.

Prof. Neeraj Bhargava, Abhishek Kumar, Pramod Singh, Manju Payal, December 17 Volume 3 Issue 12, "An Adaptive Analysis of Different Methodology for Face Recognition Algorithm", *International Journal on Future Revolution in Computer Science & Communication Engineering (IJFRSCE)*, pp. 209–212.

Yuille, A.L., Hallinan, P.W. & Cohen, D.S. Feature extraction from faces using deformable templates. *Int J Comput Vision* 8, 99–111 (1992). https://doi.org/10.1007/BF00127169.

Z. Lei, C. Wang, Q. Wang and Y. Huang, "Real-Time Face Detection and Recognition for Video Surveillance Applications," *2009 WRI World Congress on Computer Science and Information Engineering, Los Angeles, CA*, 2009, pp. 168–172.

9

Intelligent Robot for Automatic Detection of Defects in Pre-Stressed Multi-Strand Wires and Medical Gas Pipe Line System Using ANN and IoT

S K Rajesh Kanna[1]*, O. Pandithurai[1], N. Anand[2], P. Sethuramalingam[2] and Abdul Munaf[2]

[1]Professor, Rajalakshmi Institute of Technology, Chennai, India
[2]Assistant Professor, Rajalakshmi Institute of Technology, Chennai, India

Abstract

A robot with vision inspection has been developed for detecting defects on the multi-strand wires in long-span cables. The developed system consists of a climbing robot, camera for image capturing, IoT modules for transmitting images to the cloud, image processing platform, and an artificial neural network module intended for decision making. The climbing robot holds the cable with the grooved wheels along with the auto trigger camera and the IoT module. For inspection, the robot ascends along the cables continuously and acquires images of various segments of the cable. Then the captured images have been sent to the cloud storage through the IoT system. The stored images have been retrieved and their sizes have been reduced through the image processing techniques. The reduced image data have been provided as input response to artificial neural network module for decision making about the defect identification. The obtained experimental results demonstrated and proved that the projected intelligent vision robot inspection technologies are the best fit for inspection and condition assessment of pre-stressed multi-strand cables. The developed robot can also be used to inspect the medical gas pipeline systems. The robot has been programmed to detect the colors of O_2, N_2, vacuum and air pressure pipe lines.

Keywords: Inspection, vision system, ANN, Robot, pre-stressed multi-strand cable

*Corresponding author: skrkanna@gmail.com

Neeraj Bhargava, Ritu Bhargava, Pramod Singh Rathore, and Rashmi Agrawal (eds.) *Artificial Intelligence and Data Mining Approaches in Security Frameworks,* (155–172) © 2021 Scrivener Publishing LLC

9.1   Introduction

Due to globalization and advancement in automobile industries, transportation industries and logistics, road transportation became a crucial part of everyday life, both in domestic and industrial domains. So bridges are constructed over rivers and valleys for quick transportation and to connect the various segments of the countries. Thus bridges became a vital part of present transportation to connect all parts of the countries. Larger road traffic volumes in the developing countries and extreme environments led to deterioration of the bridge structures. So a necessity arises for the periodic structural monitoring and maintenance for the bridges (D. Wang, 2017). In the cable stayed bridges, pre-stressed multi-strand cables are the critical load-carrying parts requiring higher monitoring and maintenance. These pre-stressed multi-strand cables are composed of bunches of high-strength steel strands, twisted over one another to create permanent prestress on the cables; thereby the cable will not bendeformed due to major loads. Any damage in the wires or strands or any parts in the cable will progress at a faster rate and thus become a greater threat to the safety of the bridges (A. Malekjafarian, 2015). The cable materials are easily damaged, as it is in the pre-stressed condition and in the high humidity sector. So the cable damages have to be detected in time and it is essential to apprehend the present state, along with the probable forthcoming service lifespan to avoid calamities (Anand Nagarajan, 2017).

In some of the bridges, steel wrappings or the PVC closers have been used to cover and firmly grip the wire strands. At present, non-destructive inspection methods like X-rays, magnetic flux, gamma rays, eddy current, ultrasonic, etc., has been done in detecting damage in the cables. Even though the methods are producing satisfactory results, sometimes barriers have been caused in the interior of the steel wrapping for the signals, which may lead to incorrect results (C. Gao, X.K. Li, Y.C. Guo, *et al.*, 2011). Moreover, the outer circumferences of the multi-strand cables are too large and thereby prevent the non-destructive inspection methods.

In some cases, polyethylene pipes are used as the protective layer for the bridge cables. Due to ageing or accidents, if any any cracks are formed in the protective pipes, corrosive substances will start to penetrate into internal multi-strand wires, and is a challenging task to detect the damage (E. Nazarian, 2016). Laser scanning methods are used for efficient damage detection, but the system is comparatively very large, difficult to handle and more expensive (E.Q. Wu, 2009). The image of the cable bridge is shown in Figure 9.1.

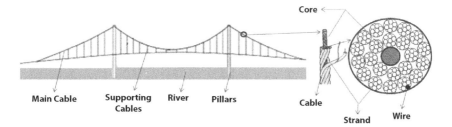

Figure 9.1 Cable bridge.

Thus, a necessity arises to develop a more competent automatic inspection model to detect surface defects for the bridge cables at low cost and with less maintenance. So in this research, machine vision technique has been applied to inspect the surface of the cable for the detection of damage. As the cable is cylindrical in shape, two cameras opposite to each other have been fixed over the rope climbing robot to cover the entire circumference of the bridge cable. Both the cameras have been triggered simultaneously to acquire images of the surface of the cables. In order to enhance the performance of the movement of the robot over the cables, a need arises to reduce the weight carried by the robot, so the captured images have been sent to the cloud through IoT Node MCU board for further processing. The cloud storage acts as the database for the inspection and interpretation for various decision-making processes in the future also. The major challenges faced in the machine vision inspection system are the variations in the images with different lighting conditions, dents in the cover portion, rust and dust over the cables, etc. Further, a need arises for the huge number of template images for comparison and the computation time also higher, which might not be the best method for the dynamically moving environments. So in this research, artificial neural networks have been utilized for judgment making, which can accommodate the deviations in the images like scale formations, rotation, reflection, brightness and smaller variations in the images due to denting and dust (F.Y. Xu, X.S., 2011). The developed system has been validated and it was found that the developed intelligent module can identify the defected portions of the bridge cable effectively and it is suitable for real-time commercial inspections. Furthermore, the computational complexities have been reduced significantly by using the SPIHT method for image compression without data loss.

Further, the developed robot has been used to inspect the medical gas pipeline system (MGPS) for defects and any gas leakages. The various gases commonly used for the MGPS are Oxygen (O2), Nitrous oxide (N2O),

Medical air 400 KPa or 4 bar / 700 KPa or 7 bar, Carbon dioxide (CO2), Nitrogen (N2), and Medical vacuum. These gases are used to sustain the life of patients through ventilators, anesthesia machine, respiratory applications, surgical equipment, insufflations purpose, scavenging system, etc. These pipes are laid outside and may be exposed to many hazardous environments. So the periodic inspection is essential, as it is used for lifesaving health care applications.

9.2   Inspection System for Defect Detection

In general, the bridge cables are exposed to high corrosion, high temperature, humidity, heavy load, wind, dust and eroded particles settlement, high stress, aging, rainfall, etc. So the cables are designed to withstand these challenges through usage of Galvanized steel wires, packaged pipes over the cables, etc. Any defects over the covering pipe or the cables will pave the way for easy propagation of that defect all over the cable. So a necessity arises to detect the defects in advance. Among various methods for detecting the defects, in this research work a machine vision system has been adopted and successfully implemented with satisfactory performances with the climbing robot (H.N. Ho, K.D. Kim, Y.S. Park, *et al.*, 2013). The inspection system for the defect detection over the bridge cable have been categorized into five different submodules such as (1) Machine vision module, (2) IoT module, (3) Image processing module, (4) ANN module and (5) Robot module in this research work.

In the machine vision module, two CCD cameras have been employed to acquire images of the surface of the bridge cable. These two cameras are placed opposite to each other and can cover the entire circumference of the cable. These cameras are triggered simultaneously for every 10 seconds or 50 cm movement of the robot. The triggering time depends on the speed and movement of the developed robot (J. Zhang, F.T.K., 2013). For every triggering the camera captures the 50 cm of the cable image and it will depend on the angle and the focal length of the camera (J.J. Lee, K.D. Kim, 2012). LEDs are used as the light source for the camera and the same led have been used to notice the location of the robot by users while travelling over the cable. The next stage followed by image acquisition is image processing, in order to diminish the weight of the system. In this research, image processing has been carried out in the remote sector, by storing the images in the cloud database. The second stage is the IoT module in which the Internet of Things (IoT) facilitates the storage and retrieval of the images from the cloud database. The node MCU board with Wi-Fi module

ESP 8266 has been used to transfer the images to the cloud and is shown in Figure 9.2a. It is configured with 32-bit low powered CPU and Wi-Fi standards. Image signals from the camera are sent to the board and those images can be given to the cloud database (Jayakumar, Rajesh Kanna S.K, 2010). The images are stored in the database with the date and time. The IoT circuit board is powered with the 12v battery in the robot.

HEADER	HEADER INFORMATION
	HEADER IDENTIFIER = 19778,
	FILE SIZE = 57174,
INFO HEADER	REVERSED 1 = 0,
	REVERSED 2 = 0,
OPTIONAL	OFFSET TO IMAGE DATA = 54,
	INFO HEADER
	INFO HEADER SIZE = 40,
PALETTE	WIDTH OF IMAGE = 159,
	HEIGHT OF IMAGE = 119,
IMAGE DATA	NO. OF COLOR PLANES = 1,
	BITS PER PIXEL = 24,
	COMPRESSION DETAILS = 0,
	SIZE OF THE IMAGE = 57120,
	PIXELS PER METER IN X AXIS = 0,
	PIXELS PER METER IN Y AXIS = 0,
	NO. OF COLORS USED<NOT FOR 24 BIT IMAGE> = 0,
	NO. OF IMPORTANT COLORS = 0,

IMAGE INFORMATION
202, 198, 198, 200, 195, 196, 198, 194, 192, 205, 201, 199, 214, 207, 206, 210, 204, 203, 207, 201, 1, 203, 220, 212, 20, 60, 59, 48, 63, 60, 49, 63, 60, 49, 63, 62, 51, 63, 62, 51, 64, 63, 52, 64, 63, 52, 65, 64, 53, 65, 64, 53, 6, 63, 54, 65, 63, 54, 0, 200, 199, 190, 200, 199, 190, 200, 199, 190, 200, 199, 190, 200, 199, 190, 200, 199, 190, 200, 19999, 190, 200, 199, 2, 205, 217, 217, 208, 203, 204, 194, 187, 190, 180, 129, 132, 123, 78, 71, 72, 71, 64, 65, 74, 54, 66, 6, 61, 62, 67, 62, 63

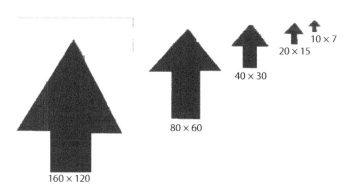

Figure 9.2 Wifi board, 2b Bitmap format of sample image, 2c SPIHT method.

The third module is the image processing module. In this stage, the acquired images have to filter for the noises and have to reduce its size to the minimum without data loss, so the processing speed of the ANN increases and computation time isreduced. In particular, the developed system reads the input images from the cloud and processes it to revert back to ANN as input. The algorithms were transcribed in Visual Basic programming language. The module reads the image from the cloud database and converts the color bitmap data into monochrome bitmap data by considering 0 and 1 for the pixel value less than 150 and greater than 150 respectively. The sample de facto format read by the software is shown in Figure 9.2b. From the 54th bit of image data section, each pixel of the image has been represented by three color intensities values such as Blue, Red and Green, and the average of the value gives the monochrome data of the pixels. To clean the noise, every pixel has been extracted and compared to the surrounding eight pixels, then the abnormalities have been removed, as is shown in Figure 9.3. Again, a large quantity of raster data in the image has been reduced to smaller size using SPIHT method. In this method, filtered image has to be divided into four equal quadrants. Yet again every single quadrant has been divided into four equal quadrants and so on (M.R. Kaloop, J.W., 2015). In the final matrix, the maximum repetitive code will be anticipated as value for that matrix and a sample reduction of size is shown in Figure 9.2c (N. Imajo 2015). In this approach, the size has been reduced to 70 bits from 786432 bits. This 70-bit data have been given as input to ANN.

From Figure 9.3, it is clear that the noise level in the captured images has been cleared.

The methodology of the developed model is shown in Figure 9.4. The fifth stage is the artificial neural network model, for experimentation, ANN have formulateed with 70,125, 80 and 21 neurons in input, hidden and output layer, respectively. In this research, expected mean squared error value has

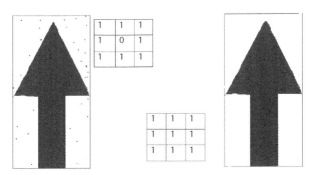

Figure 9.3 Noise reduction.

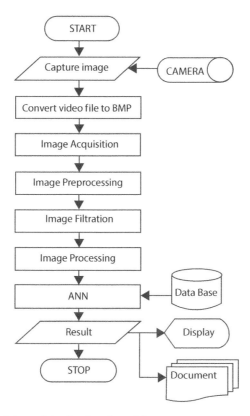

Figure 9.4 Methodology of the developed model.

been set to 0.01, total output error, i.e., T error = $\sum$ Dop – Oop, is set to 0.01, learning rate is set as 0.3 (R. Christen, 2009). The developed network is given in Figure 9.5. The fifth stage is the cable climbing robot development stage. It has a carriage module to carry the cameras, battery, Wi-Fi and control board.

Figure 9.5 shows the robot carriage and frame drawing. The entire carriage has been fixed with the frame, which has grooved wheels for movements. The V-grooved wheels have been provided with individual actuation motors to move all over the cable in all the directions along with the compression springs to hold the wheels with the cable, and they can roll over the cable without much slippage. Groove size of the wheels has to be set based on the diameter of the cable, for superior performance. Wheels size considered for the experimentation have 150 mm diameter. The developed robot have been studied with its dynamics for the balancing, thereby the performance of the climbing enhanced (S K Rajesh Kanna, 2017).

In the first stage, the image is captured using digital camera as bitmap image. In the processing stage, the color bitmap data is converted into

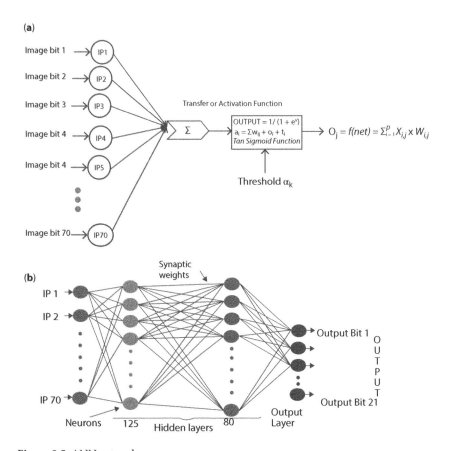

Figure 9.5 ANN network.

monochrome bitmap data. The noises in the data are filtered using filter algorithm. In the processing stage the size of image data is converted into the one acceptable by the trained BPN algorithm using SPIHT. This processed data is sent to ANN for image inspection. Then the output of ANN is interpreted to inspection of the component. These stages are explained in detail in the following sections (S K Rajesh Kanna, 2018).

9.3   Defect Recognition Methodology

The essential component for the machine vision system is the light source; LED has been used as the light source as its brightness can be easily controlled. Also the LED is used as the indicator for locating the robot over the high-altitude cables. CCD cameras are the core parts for the damage detection system. When the climbing robot crawls over the bridge cable, for every 50 cm movement, the Arduino controller sends a trigger signal to the camera. Immediately after receiving the command, the cameras capture the surface image of the cable and the acquired surface images send to the cloud database through the IoT module. The cameras are placed exactly opposite to each other and at the same focal length. The single frame of the camera image is of 1024 x 768 pixel resolution and is stored in the bitmap format for easy processing. The images are uploaded to the cloud and the cloud updates the data at the predefined time.

Then the developed image processing module processes the cloud-stored images at remote location. The various stages of image processing are the cropping of the edges, filtering of the noisy data, conversion to mono chrome images to reduce 1/3 of its size, conversion to binary image for easy processing by ANN and size reduction without data loss or the shape loss using set partitioning in Hierarchical tree method to 28 times. The final size of the image contains 70 binary digits (S.K. Rajesh Kanna, 2012).

These 70 binary digits have been given as input to the developed trained ANN module. The number of neuron in the input layer is 70, followed by two hidden layers of 125 and 80 neurons and an output layer with 21 neurons. The sample image with ANN encoded sequence and the decisions for the ANN sequence are shown in Figure 9.7 (V. Torres, 2011).

Before implementation, the sample 150 images of the cables and the cover pipes have been collected, 125 images have been used to train the artificial neural network and 25 images have been used to test the network. For training the network back propagation algorithm with the tan sigmoid function have been used. The back propagation algorithm have modified the threshold values of the neurons and the weight values of the links from the back layer to the front layer based on the obtained error value (Vignesh, Rajesh Kanna S K, 2017).

The equations and the formulas used for the updation are given in Figure 9.4. Initially the threshold and the weight values are assumed to the random values. The final trained weight and the threshold values are stored in the notepad file and the same again stored in the cloud. This trained ANN module processes the given data and values obtained at the output layers have to be encoded to a human understandable format (N. Bhargava, *et al.*, 2017). The sample combination of the output and the decision values have been shown in Figure 9.6. For the various combination of 21 binary digits, the decoded decisions are no image, unknown image, no defect, minor defect or dust, defected cable and rusted cable. The various types of sample images and the decisions are shown in Figure 9.7.

The experimentation have been conducted in the laboratory with the different types of cables and the covered pipes at different angles and the

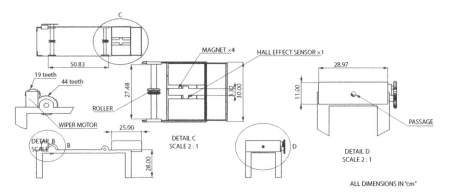

Figure 9.6 Robot carriage and frame.

orientations. The results provided by the robot are quite satisfactory. So the developed modules can be used for the inspection of the bridge cable. The major challenges faced in the implementations are the arbitrary climbing speed of the robot at various orientations, periodic uploading of images to the cloud, and computational time is slower than the capturing time, so the immediate prediction cannot be possible (W.R. Wickramasinghe, 2016).

The developed model can be used for other applications also. so the developed model is used for the health care applications.

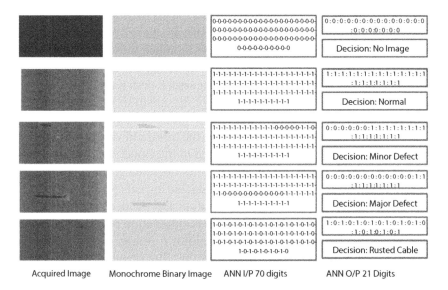

Figure 9.7 Sample images and ANN decisions.

9.4   Health Care MGPS Inspection

In most hospitals, MGPS are commonly used for various lifesaving activities. Also these pipes are under high pressure and exposed to outside environments. The sample piping system is shown in Figure 9.8.

The developed robot can be used to detect the defects in the piping system periodically. Normally the pipes used for the MGPS are copper pipes and are of different colors for easy identification. The colors and the gases inside the pipe are given in Figure 9.8. The robot has been programmed for the color codes provided in Figure 9.9.

Figure 9.8 Sample MGPS.

GAS	ISO Color Code	
Carbon Dioxide		Grey
He-O$_2$		Brown & White
Instrument Air		
Medical Air		Black & White
Nitrogen		Black
Nitrous Oxide		Blue
O$_2$-He		White & Brown
Oxygen		White
Vacuum (Suction)		Yellow
WAGD (Evac)		Purple

Figure 9.9 MGPS – color codes.

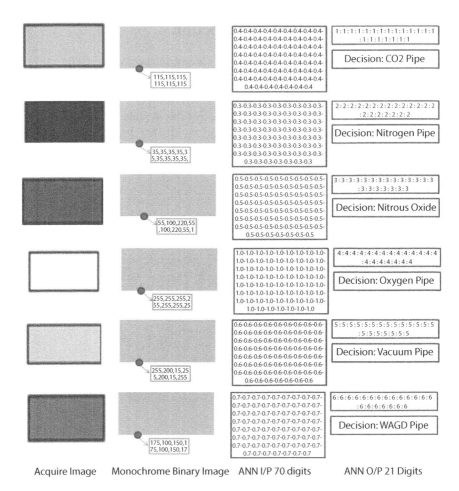

Acquire Image   Monochrome Binary Image   ANN I/P 70 digits   ANN O/P 21 Digits

Figure 9.10 Image and ANN encoding for color pipes.

The developed robot has been programmed to operate in three different modules. In the first module, based on the color code over the gas pipe, the robot detects the type of gas inside it and informs the user (Bhargava, A., 2017). In the second module, in the corresponding gas pipe any damages or defects will be identified. In some of the hospitals, for economic reasons, the color codes are given only in the patient's room, outside is with the copper tubes.

So in the third module, the developed robot can detect the defects over the copper tubes. The various acquired images and the detection mechanisms are given in Figure 9.10 and Figure 9.12. Figure 9.10 describes the various color codes and the ANN input and output for the different gas pipes. Figure 9.11 denotes the defects in the color coded pipes. Figure 9.12 denotes the copper pipe and the encoded digit representation.

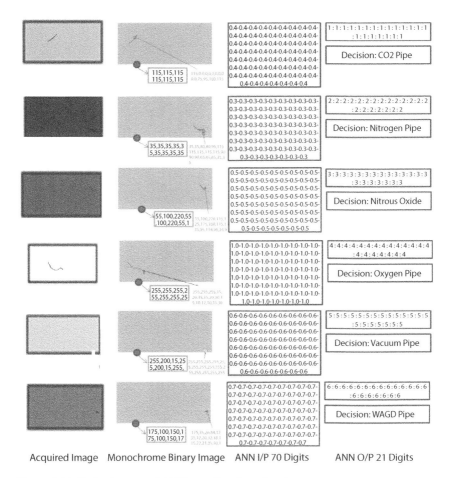

Figure 9.11 Defect image and ANN encoding for color pipes.

Figure 9.12 Copper pipe image and ANN encoding.

Even though the robots detect the defects in the pipe lines, leakages are also detected by having the fan and the generator mechanism. Any leakage in the gas will rotate the fan blades and thereby the generator generates the emf; once the emf is read by the microcontroller, the warning signal is given to the user.

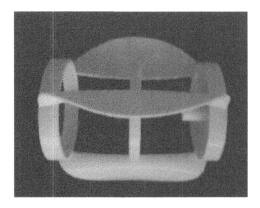

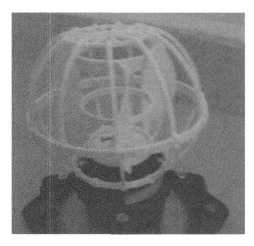

Figure 9.13 Leakage deduction fan.

The images of the developed fan blade for detecting the gas leakage are shown in Figure 9.13. The blades are made using 3D printer and the special properties are lightweight and can rotate for the gas hit on any area of the surface of the blades. The fan blades differ from the commercial blades.

9.5   Conclusion

In this work, a vision-based intelligent inspection robot for bridge cable inspection has been developed with IoT technologies for storing the data in the cloud for processing. The robot has been employed to detect surface damage on the bridge cables. The surface defects over the bridge cables can be automatically detected and intimated to the remote system for the necessary maintenance actions. Meanwhile, the simplified SPIHT can be efficiently implemented for the size reduction without much data loss. Investigational outcomes concluded that the image acquisition, image processing and defect identification have been accomplished efficiently for bridge cable/rope surface damage detection by the developed model. Future work can be on real-time processing, enhanced robot movement over the cable and parallel processing to reduce computation time. Also, the developed robot has been used to test the medical gas pipe line system for defects and leakages.

References

D. Wang, H. Lu, M.H. Yang, Robust visual tracking via least soft-threshold squares, i. 26 (9) (2016) 1709–1721.

Singh Rathore, P., Kumar, A., & Gracia-Diaz, V. (2020). A Holistic Methodology for Improved RFID Network Lifetime by Advanced Cluster Head Selection Using Dragonfly Algorithm. *International Journal of Interactive Multimedia and Artificial Intelligence*, 6 (Regular Issue), 8. http://doi.org/10.9781/ijimai.2020.05.003

A. Malekjafarian, P.J. McGetrick, E.J. Obrien, A review of indirect bridge monitoring using passing vehicles, *Shock. Vib.* (2015) 1–16, Article ID 286139.

Anand Nagarajan, S. K. Rajesh Kanna, V. Manoj Kumar, "Multibody dynamic simulation of a hyper redundant robotic manipulator using ADAMS ansys interaction", *IEEE Explorer*, 2017, DOI: 10.1109/ICAMMAET.2017.8186629.

C. Gao, X.K. Li, Y.C. Guo, *et al.*, Cable health monitoring system for long-span cable-stayed Bridge, *Disaster. Adv.* 4 (2011) 130–135.

E. Nazarian, F. Ansari, X. Zhang, *et al.*, Detection of tension loss in cables of cable-stayed bridges by distributed monitoring of bridge deck strains, *J. Struct. Eng.* 142 (6) (2016) 1–13.

E.Q. Wu, Y.L. Ke, B.J. Du, Noncontact laser inspection based on a PSD for the inner surface of minidiameter pipes, *IEEE Trans. Instrum. Meas.* 58 (7) (2009) 2169–2173.

F.Y. Xu, X.S. Wang, L. Wang, Cable inspection robot for cable-stayed bridges: design, analysis, and application, *J. Field Robot.* 28 (3) (2011) 441–459.

H.N. Ho, K.D. Kim, Y.S. Park, *et al.*, An efficient image-based damage detection for cable surface in cable-stayed bridges, *NDT&E Int.* 58 (2013) 18–23.

P. S. Rathore, A. Chaudhary and B. Singh, "Route planning via facilities in time dependent network," *2013 IEEE Conference on Information & Communication Technologies*, Thuckalay, Tamil Nadu, India, 2013, pp. 652-655. doi: 10.1109/CICT.2013.6558175

N. Bhargava, S. Dayma, A. Kumar and P. Singh, "An approach for classification using simple CART algorithm in WEKA," *2017 11th International Conference on Intelligent Systems and Control (ISCO)*, Coimbatore, 2017, pp. 212-216. doi: 10.1109/ISCO. 2017. 7855983

Naveen Kumar, Prakarti Triwedi, Pramod Singh Rathore, "An Adaptive Approach for image adaptive watermarking using Elliptical curve cryptography (ECC)", *First International Conference on Information Technology and Knowledge Management*, pp. 89–92, ISSN 2300-5963 ACSIS, Vol. 14 DOI: 10.15439/2018KM19

J. Zhang, F.T.K. Au, Effect of baseline calibration on assessment of long-term performance of cable-stayed bridges, *Eng. Fail. Anal.* 35 (2013) 234–246.

J.J. Lee, K.D. Kim, H.N. Ho, A vision-based damage detection of cable exterior in cable-stayed bridges, in: *Sixth International Conference on Bridge Maintenance, Safety and Management, Stresa, Lake Maggiore, Italy, 2012*, pp. 1138–1145.

Jayakumar, Rajesh Kanna S.K., "Inspection System for Detecting Defects in a Transistor Using Artificial Neural Network (ANN)", *IEEE Explorer*, INSPEC Accession Number: 11887499, Dec. 2010, pp. 76 – 81.

Bharat Singh, Ravinder Singh and Pramod Singh Rathore. Article: Randomized Virtual Scanning Technique for Road Network. *International Journal of Computer Applications* 77(16):1-4, September 2013.

M.R. Kaloop, J.W. Hu, Stayed-Cable bridge damage detection and localization based on accelerometer health monitoring measurements, *Shock. Vib.* (2015) 1–11, Article ID 102680.

N. Imajo, Y. Takada, M. Kashinoki, Development and evaluation of compact robot imitating hermit crab for inspecting the outer surface of pipes, *J. Robot.* 2015 (2015) 1–7.

R. Christen, A. Bergamini, M. Motavalli, Influence of steel wrapping on magneto-inductive testing of the main cables of suspension bridges, *NDT&E Int.* 42 (2009) 22–27.

Prof. Neeraj Bhargava, Abhishek Kumar, Pramod Singh, Manju Payal, December 17, Volume 3 Issue 12, "An Adaptive Analysis of Different Methodology for Face Recognition Algorithm", *International Journal on Future Revolution in Computer Science & Communication Engineering (IJFRSCE)*, pp: 209 – 212.

N. Bhargava, S. Sharma, R. Purohit and P. S. Rathore, "Prediction of recurrence cancer using J48 algorithm," *2017 2nd International Conference on Communication and Electronics Systems (ICCES)*, Coimbatore, 2017, pp. 386-390, doi: 10.1109/CESYS.2017.8321306.

S K Rajesh Kanna, G Muthu and A. Venkatesan, "Inspection of Boiler Pipes Using Miniature Mobile Robot", *International Journal of Advanced Research in Management, Engineering and Technology*, 2017, vol. 2, issue 4, pp. 677-680.

Pramod Singh Rathore, "An adaptive method for Edge Preserving Denoising "International Conference on Communication and Electronics Systems, Institute of Electrical and Electronics Engineers" & PPG Institute of

Technology,. (2017). *Proceedings of the 2nd International Conference on Communication and Electronics Systems (ICCES 2017)*: 19-20, October 2017.

S K Rajesh Kanna, Lingaraj .N, Muthuvel. C, Mohan. D, "ANN for Profit Predication in Maize Crop Cultivation through Cloud Database", *International Journal of Engineering and Science Invention*, 2018, vol. 7, issue 2, PP. 63-70.

S.K. Rajesh Kanna, and M. Saravana Manigandan, "Intelligent Vision Inspection System for IC Engine Head: An ANN Approach", *Journal of Advanced Material Research*, Trans tek publications, Feb 2012, Vols. 479-481 (2012), pp. 2242 - 2245.

S.K. Rajesh Kanna, N. Anand, R. Mohanraj, "Intelligent Vision Based Pneumatic Wall Painting Machine: an ANN Approach", *International Journal of Engineering and Applied Sciences*, ISSN 2305-8269, August, 2015.

V. Torres, S. Quek, B. Fernandes, *et al.*, Development of a scanning system to detect corrosion in wire bundles and pipes using magnetoresistive sensors, *Insight* 53 (2) (2011) 82–84.

Prof. Neeraj Bhargava, Pramod Singh, Abhishek Kumar, Taruna Sharma, Priya Meena, December 17 Volume 3 Issue 12, "An Adaptive Approach for Eigenfaces-based Facial Recognition", *International Journal on Future Revolution in Computer Science & Communication Engineering (IJFRSCE)*, pp: 213-216.

Vignesh, Rajesh Kanna S K, N Lingaraj, "Intelligent Automated Guided Vehicle Using Visual Servoing", *American Journal of Engineering Research*, 2017, vol. 6, issue 11, pp. 16-20.N.

Bhargava, A. Kumar Sharma, A. Kumar and P. S. Rathoe, "An adaptive method for edge preserving denoising," *2017 2nd International Conference on Communication and Electronics Systems (ICCES)*, Coimbatore, 2017, pp. 600-604, doi: 10.1109/CESYS.2017.8321149.

W.R. Wickramasinghe, D.P. Thambiratnam, T.H.T. Chan, *et al.*, Vibration characteristics and damage detection in a suspension bridge, *J. Sound Vib.* 375 (2016) 254–274.

10

Fuzzy Approach for Designing Security Framework

Kapil Chauhan

Aryabhatt Engineering College, Ajmer, India

Abstract

Cyber security isn't just one difficulty; typically it's a problem involving many different aspects. Fuzzy Rule–based object/system for cyber security may be a system that consists of pools of rules and a mechanism for accessing and running the principles. The pools of rules are typically constructed with a set of related rule cluster/set. The impact of cyber criminal activity depends on the character of the crime and the nature of the victim. Saudi Arabia faces numerous cyber threats including Denial of Service (DoS), malware, website defamation, and spam and phishing email attacks. Although recent findings highlight the poor state of Saudi Arabia's information security system, it's within the present premise's suggestion that a special cyber security risk assessment can be made. Using Fuzzy logic theory, we propose Fuzzy Inference Model (FIS) to supply risk mitigation and check out to unravel such issues to proposed entities.

The objective of interruption location is to watch organize exercises and consequently, recognize vindictive assaults and decide a right design of the PC arrange security.

Keywords: Fuzzy logic, software development, cyber security, fuzzy rules, malicious threats, fuzzy inference system, software security, security patterns

10.1 Introduction

Today associations, looking with a decent scope of likely Threats to their data security (IS), are progressively inquisitive about a significant level of it. One among the least difficult approaches to gauge, accomplish and

Email: kapilajmer86@gmail.com

Neeraj Bhargava, Ritu Bhargava, Pramod Singh Rathore, and Rashmi Agrawal (eds.) Artificial Intelligence and Data Mining Approaches in Security Frameworks, (173–196) © 2021 Scrivener Publishing LLC

keep up security of information is an Information Security review. Review of security (comprehensively checked) might be a mind boggling, many-stage work and serious procedure involving highly qualified professionals (specialists) in IS, which makes it very costly.

Vulnerability influences dynamic; the idea of information is innately identified with the idea of vulnerability. The first principal part of this association is that the vulnerability engaged with any critical thinking circumstance might be a consequences of some data insufficiency, which can be inadequate, uncertain, fragmentary, not completely solid, ambiguous, opposing, or lacking in another way. Vulnerability is a trait of information.

Due to IoT, in future individuals will have an undetectable and omnipresent processing foundation to perform various exercises both at work and at home. The present-day home requires simple to utilize and synergistic gadgets. At first representative rationale control was acquainted with model free control configuration approach yet was censured due to the absence of efficient security investigation and regulator plan.

1. Scoping and pre-review study: deciding the most zone of center; setting up review destinations.
2. Arranging and readiness: for the most part producing a review work plan/agenda.
3. Hands on work: gathering proof by meeting staff and administrators, assessing archives, printouts and information, watching forms in real life, and so on.
4. Investigation: looking at, checking on and analyzing of the amassed proof concerning the goals.
5. Announcing: exploring every single past stage, discovering relations inside the gathered data and creating a report.
6. Conclusion. Every one of the levels has an outsized measure of data, which must be recorded, sorted out and, at last, broken down.

The quick advancement of information innovation, system and PC assaults have ignited wide concern around the world. There has not just been an increase in the number and kind of assaults, the multifaceted nature and class has additionally been expanded. The possible damages of assaults are progressively genuine. As Internet security might be a quick-moving field, the assaults that are getting the features can change fundamentally from one year to the next.

A Fuzzy Expert System (FES) upheld estimated thinking is characterized and depended on to work out reasonable competitor transports during a dispersion framework for capacitor situation. Voltages and force misfortune decrease records of dissemination framework transports are demonstrated by Fuzzy enrollment capacities. A Fuzzy Expert System containing a gathering of heuristic standards is then utilized to work out the capacitor arrangement appropriateness of each transport inside the conveyance framework. Capacitors are then positioned on the hubs with the absolute best reasonableness utilizing the ensuing advances:

- A load stream program figures the force misfortune decrease of the framework regarding each transport when the receptive burden current at that transport is redressed. The PLRs are then straightly standardized into a [0,1] territory with the absolute best misfortune decrease being 1 and in this manner the most minimal being 0.
- The power-misfortune decrease records close by the per-unit nodal voltages (before PLR pay) are the two data sources that are utilized by the Fuzzy Expert System. The FES at that point decides the premier appropriate hub for capacitor establishment by Fuzzy deduction.
- Investment funds work concerning the yearly expense of the capacitor (controlled by size) is characterized and in this way the reserve funds accomplished from power-misfortune decrease is amplified to work out the ideal estimating of capacitor bank to be put at that hub.
- The above procedure will be rehashed, finding another hub for capacitor position, until not any more budgetary reserve funds are frequently figured it out.

1. Arranging and readiness: normally creating a review workplan/agenda.
2. Hands-on work: gathering proof by talking to staff and supervisors, looking into reports, printouts and information, watching forms in real life, and so on.
3. Examination: looking at, checking on and analyzing of the gathered proof regarding the goals.
4. Detailing: inspecting every single past stage, discovering relations inside the gathered data and creating a report.

Fuzzy rationale might be such a many-esteemed rationale or probabilistic rationale; it manages thinking that is estimated rather than fixed and definite. Interestingly with customary rationale they will have differing values, where double sets have two-esteemed rationale, valid or bogus, emblematic rationale factors may have a reality esteem that ranges in degree somewhere in the range of 0 and 1. Emblematic rationale has been stretched out to deal with the idea of fractional truth, where the truth worth may range between totally obvious and totally bogus. Besides, when phonetic factors are utilized, these degrees could likewise be overseen by explicit capacities. Representative rationale strategies have been utilized inside the PC security field since the 1990s. Representative rationale includes additionally showed potential inside the interruption discovery field in contrast with frameworks utilizing severe mark coordinating or exemplary example deviation recognition. The idea of fluffiness assists with smoothing the sudden partition of typical conduct from irregular conduct. Representative rationale includes a capacity to speak to loose kinds of thinking in territories where firm choices should be made in inconclusive situations like interruption recognition. Fuzzy expert system should be describe in Figure 10.1 (R. Chandia *et al.*, 2007).

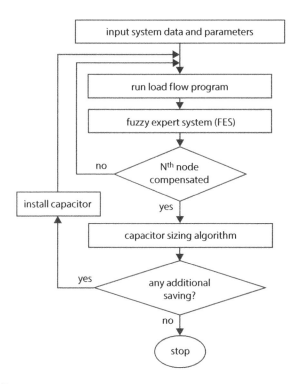

Figure 10.1 Fuzzy expert system.

10.2   Fuzzy Set

1. Fuzzy set might be a set having degrees of participation somewhere in the range of 1 and 0. Fuzzy sets are spoken to with tilde character (~). For example, the number of vehicles following traffic lights at a particular excursion of all vehicles present will have participation esteem between [0,1].
2. Halfway participation exists when individual from 1 Fuzzy set likewise can be an area of other Fuzzy sets inside a similar universe.
3. The level of participation or truth isn't the same as likelihood. Fuzzy truth speaks to enrollment in ambiguously characterized sets.

Prologue to Fuzzy Set during this part, the idea of Fuzzy sets and consequently the procedure on the Fuzzy set are talked about. The ideas are the speculations of fresh sets. Old style sets likewise are called 'fresh' sets so on recognize them from Fuzzy sets. Truth be told, the Crisp sets are frequently taken as unique instances of Fuzzy sets. Leave An alone a fresh set characterized over the Universe X. At that point for any component x in X, either x might be an individual from An or not. In Fuzzy unadulterated science, this property is summed up. Accordingly, during a Fuzzy set, it's a bit much that x might be a full Member of the set or not a part. It is frequently a halfway individual from the sets mention in Figure 10.2.

The speculation is proceeded as follows: For any fresh set An, it's conceivable to characterize a Characteristic capacity or enrollment work $\mu P = \{0, 1\}$. i.e., the trademark work takes both of the qualities 0 or1 inside the

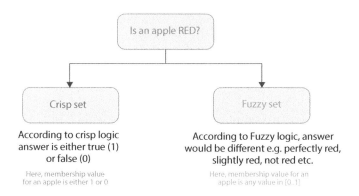

Figure 10.2 Crisp vs. fuzzy sets.

traditional set. For a Fuzzy set, the trademark capacity can take any an incentive somewhere in the range of zero and one.

Definition:
The enrollment work μP(x1) of a Fuzzy set A might be a capacity μP : X1→ ['true' , 'false'] So, every component in x1 in X has participation degree: μA(x1) ∈ ['true' , 'false'] An is completely controlled by the arrangement of tuples: P = {(x1, μP(y)) x∈ X}.

Example:
Assume somebody needs to clarify the object of vehicles having the property of being costly by considering BMW, Rolls Royce, Mercedes, Ferrari, Fiat, Honda and Renault. A few vehicles like Ferrari and Rolls Royce are unquestionably costly and a couple, like Fiat and Renault, aren't costly when contrasted and don't have a place with the set. Utilizing a Fuzzy set, the Fuzzy arrangement of pricey vehicles are frequently portrayed as:

The Fuzzy set is undifferentiated from the very arrangement of the Boolean rationale with additional participation capacities in the middle of "valid" and "bogus". As its name suggests, it's the rationale basic methods of thinking which are estimated rather than definite. The significance of emblematic rationale gets from the very actuality that the larger part methods of human thinking and especially sense thinking are estimated in nature.

The fundamental attributes of emblematic rationale are as per the following:

- In emblematic rationale, accurate thinking is seen as a constraining instance of estimated thinking.
- In emblematic rationale everything might involve degree.
- Any rationale are regularly fuzzified.
- In emblematic rationale, information is deciphered as a lot of versatile or, identically, Fuzzy imperative on a lot of factors.
- Inference is seen as a procedure of proliferation of versatile limitations.
- There are two significant classes of crime with PCs:
- Unauthorized utilization of a PC, which could include taking a username and secret key, or may include getting to the victim's PC by means of the web through a secondary passage worked by a bug program.

- Creating or delivering a malignant PC infection (e.g., PC infection, worm, Trojan horse).

When people hear the words "System fault", they frequently consider profane pictures accessible on the web. The optimal issue on the web is typically comparable in light of the fact that the useful fact of indecency in books, aside from different logical issues with single purview on the site.

Typical faults involved System are the same from violations without Systems. The system is simply an instrument that a fraud case uses to carry out a fraud offense. Consider the following:

- utilizing a System, a scanner, designs programming, and a great shading printer for imitation or forging is that a similar fault as utilizing a good old press with ink.
- Stealing a system with exclusive data put away on the hard plate inside the system is that a similar fault as taking a satchel.
- Web administrations to request sex is practically equivalent to different types of sales, at that point isn't a substitution fault.
- System is frequently contrastingly to submit either theft or misrepresentation.

Procedure on Fuzzy sets

The notable tasks which might be performed on Fuzzy sets are the activities of association, convergence, supplement, mathematical item and arithmetical entirety. Much exploration concerning Fuzzy sets and their applications to automata hypothesis, rationale, control, game, geography, design acknowledgment, indispensable, etymology, scientific classification, framework, choosing, data recovery and so on, has been sincerely embraced by utilizing these activities for Fuzzy sets.

Notwithstanding those activities, new tasks called "limited aggregate" and moreover to those activities, new tasks called "limited entirety" and "limited distinction" are presented by Zadeh (1975) to explore the Fuzzy thinking which explains how to take care of the thinking issues which are excessively mind boggling for exact arrangement.

Different types of operators:

(i)   Balance
(ii) Supplement

(iii) Crossing point
(iv) Association
(v)   Mathematical item
(vi) Augmentation of Fuzzy set with Crisp Number
(vii) Intensity of Fuzzy set
(viii) Mathematical aggregate

1. Equivalent Fuzzy sets
   We have to consider two Fuzzy sets P(y) and Q(y) are supposed to be equivalent, if $\mu P(y) = \mu Q(y)$ for all $x \in X$. it's
   communicated as follows:

$$P(y) = Q(y), \text{ if } \mu P(y) = \mu Q(y)$$

   Note: Two Fuzzy sets P(y) and Q(y) are supposed to be
   inconsistent, if $\mu P(y) \neq \mu Q(y)$ for at least $y \in Y$.
   Example:

$$P(y) = \{(y1,0.1),(y2,0.2),(y3,0.3),(y4,0.4)\}$$
$$Q(y) = \{(y1,0.1),(y2,0.5),(y3,0.3),(y4,0.6)\}$$

   As $\mu P(y) \neq \mu Q(y)$ for different $y \in Y$, $P(y) \neq Q(y)$

2. Supplement of Fuzzy set P(y)
   The supplement is something contrary to the set. The supplement of a Fuzzy set is signified by p'(y) and is characterized as for the all inclusive set Y as follows shown in Figure
   10.3:

$$P'(y) = 1\text{-}P(y) \text{ for all } y \in Y$$

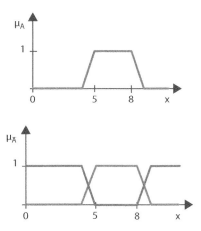

Figure 10.3 Example of complement operation on a fuzzy set.

3. Crossing points of Fuzzy sets

   Segment of a Fuzzy sets characterize what extent of the component has a place with the two sets. May have various degrees of enrollment in each set. The level of participation is that the lower enrollment in the two arrangements of each component. Let P(y) and Q(y) are two Fuzzy sets, the convergence of is meant by (P∩Q)(y) and in this manner the participation work esteem is given as follows:

$$\mu\,(P{\cap}Q)(y) = \min\{\mu P(y),\mu Q(y)\}$$

   Crossing point is practically equivalent to sensible AND activity

$$P(y) = \{(y1,0.7),(y2,0.3),(y3,0.9),(y4,0.1)\}$$
$$Q(y) = \{(y1,0.2),(y2,0.5),(y3,0.7),(y4,0.4)\}$$
$$\mu\,(P{\cap}Q)(y1) = \min\{\mu P(y1),\mu Q(y1)\} = \min\{0.7,0.2\} = 0.2$$
$$\mu\,(P{\cap}Q)(y2) = \min\{\mu P(y2),\mu Q(y2)\} = \min\{0.3,0.5\} = 0.3$$
$$\mu\,(P{\cap}Q)(y3) = \min\{\mu P(y3),\mu Q(y3)\} = \min\{0.9,0.7\} = 0.7$$
$$\mu\,(P{\cap}Q)(y4) = \min\{\mu P(y4),\mu Q(y4)\} = \min\{0.1,0.4\} = 0.1$$

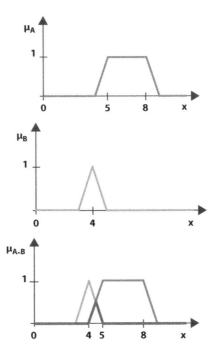

Figure 10.4 Example of intersection operation on a fuzzy set.

Diagrammatic representation of the crossing point administrator is shown in Figure 10.4.

4. Association of sets (Fuzzy)

Association of Fuzzy sets comprises each component that show into the set. The estimation of the enrollment esteem is will be the biggest participation estimation of the component in either set Let P(y) and Q(y) are two Fuzzy sets for all $y \in X$, Union of Fuzzy sets is meant by (PUQ)(y) and the enrollment work esteem is resolved as follows:

$$\mu\,(PUQ)(y) = \max\{\mu P(y), \mu Q(y)\}$$

Example:

$$P(y) = \{(y1, 0.7), (y2, 0.3), (y3, 0.9), (y4, 0.1)\}$$
$$Q(y) = \{(y1, 0.2), (y2, 0.5), (y3, 0.7), (y4, 0.4)\}$$
$$\mu\,(PUQ)(y1) = \max\{\mu P(y1), \mu Q(y1)\} = \max\{0.7, 0.2\} = 0.7$$
$$\mu\,(PUQ)(y2) = \max\{\mu P(y2), \mu Q(y2)\} = \max\{0.3, 0.5\} = 0.5$$

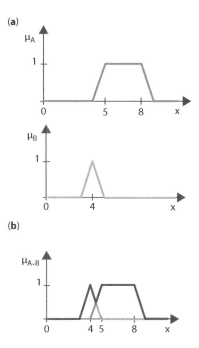

Figure 10.5 Example of union operation on a fuzzy set.

$$\mu\,(P\cup Q)(y3) = \max\{\mu P(y3),\mu Q(y3)\} = \max\{0.9,0.7\} = 0.9$$
$$\mu\,(P\cup Q)(y4) = \max\{\mu P(y4),\mu Q(y4)\} = \max\{0.1,0.4\} = 0.4$$

Note: Union is closely resembling legitimate OR activity shown in Figure 10.5.

5. Logarithmic result of sets (Fuzzy)
   The result of two Fuzzy sets P(y) and Q(y) for all $y \in Y$, is meant by P(y).Q(y)
   Characterized as given below:

$$P(y).Q(y) = \{(y,\ \mu P(y).\mu Q(y)),\ y \in Y\ \}$$

Example:

$$P(y) = \{(y1,0.1),(y2,0.2),(y3,0.3),(y4,0.4)\}$$
$$Q(y) = \{(y1,0.5),(y2,0.7),(y3,0.8),(y4,0.9)\}$$
$$P(y).Q(y') = \{(y1,0.05),(y2,0.14),(y3,0.24),(y4,0.36)\}$$

6. Product of Number (Fuzzy) by a Number (Crisp)
    The result of Fuzzy set P(y) and a Crisp Number 'd' is communicated as follows

$$P(y).Q(y) = \{(y, d . \mu P(y)), y \in Y \}$$

Example:
Let us consider a Fuzzy set P(y) to such an extent that

$$P(y) = \{(y1,0.1),(y2,0.2),(y3,0.3),(y4,0.4)\}$$
$$d = 0.2$$

at that point d.P(y) = {(y1,0.02),(y2,0.04),(y3,0.06),(y4,0.08)}

7. Intensity of a set (Fuzzy)
    The p-th intensity of a Fuzzy set P(y) yields another Fuzzy set $P^P(y)$, whose participation worth can be resolved as follows

$$\mu P^P(y) = \{\mu P(y)\}^P, y \in y \}$$
$$P > = 1, P^P(y) \text{ called (fixation)}$$
$$P < 1, P^P(y) \text{ is called (enlargement)}$$

Example:
Consider a Fuzzy set A(y)

$$P(y) = \{(y1,0.1),(y2,0.2),(y3,0.3),(y4,0.4)\}$$
$$P = 2$$

At that point $P^2(y)$ = {(y1, 0.01), (y2, 0.04), (y3, 0.09), (y4, 0.16)}

8. Mathematical whole of two sets (Fuzzy)
    In this Function, whole of two Fuzzy sets P(y) and Q(y) for all $y \in y$, is indicated by P(y)+Q(y) and characterized as follows

$$P(y) + Q(y) = \{(y, \mu P + Q(y), y \in Y\}$$

Here $\mu P + Q(y) = \mu P(y) + \mu Q(y) - \mu P(y).\mu Q(y)$
Example:

$$P(y) = \{(y1, 0.1),(y2,0.2),(y3,0.3),(y4,0.4)\}$$

$$Q(y) = \{(y1,0.5),(y2,0.7),(y3,0.8),(y4,0.9)\}$$
Presently (y) + Q(y) = \{(y1,0.55),(y2,0.76),(y3,0.86),(y4,0.94)\}

10.3   Planning for a Rule-Based Expert System for Cyber Security

Planning level incorporate characterizing digital security Expert System factors, information assortment for digital Threats, framework structure and usage. These level are depicted in the following sections.

10.3.1   Level 1: Defining Cyber Security Expert System Variables

The initial phase in the proposed model is the foundation of info and yield factors (R. Shanmugavadivu *et al.*, 2011). This undertaking is typically done by contemplating the difficult space and by counsel with the digital specialists. There is a vast number of potential applicants which ought to be confined to positive numbers.

10.3.2   Level 2: Information Gathering for Cyber Terrorism

Expert System structures the data on the human system. It gives clarifications like the human system. The information utilized for this work have been extricated from a progression of polls gathered from digital specialists and framework managers. The stored information is connected particularly with points given below Figure 10.6 (N. Fovino *et al.*, 2007).

- virus, malware, logic bomb, DoS attacks, social engineering.
- Lack of service, seizing web page, attacks for protesting, seize critical systems, capture confidential information.

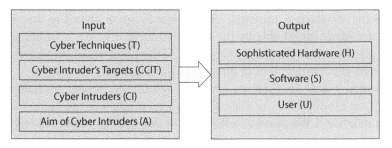

Figure 10.6 Proposed model for input and output.

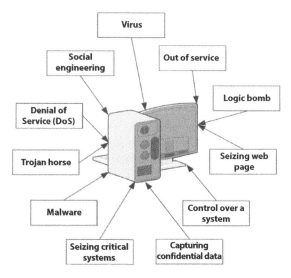

Figure 10.7 Potential cyber threats.

The analyst calculates which systems—transportation, financial hub, power system, quick services, water supply, oil and natural gas distribution stations—cyber terrorists may possibly attack describe in Figure 10.7.

10.3.3   Level 3: System Design

Knowledge-based System might be forward or in reverse anchoring. In forward association frameworks, we reason from forerunner truth to resulting truth; we reason from realities in the standard predecessor that we understand to be consistent with build up of new realities whose fact is inferred by the precursor. In reverse tying inverts this; we endeavor to search capabilities to build up reality of some optional state.

Forward Chaining: A specialist framework rule might be planned essentially as "in the event that Xn, at that point Y" where X will be a lot of circumstances on information and Y is a lot of directions to be completed when the standard is initiated. The standards are analyzed to see which rules are made initial able by the information, that is, An is fulfilled, and a standard or rules chose for executing. At the point when the standard is executed, the arrangement of directions Y is executed.

Reverse rules: An alternate grouping is followed in reverse anchoring. In reverse binding, we calculate what end we might want to come to, that is, we indicate Y. We discover a standard or decides that have the ideal subsequent, and take a look at the predecessor X to perceive what the information must be to fulfill P. Presently we discover how that information can be built up, and search for decides that have that information as a subsequent, or info information from a client to check whether the forerunner can be fulfilled. In reverse fastening we work in reverse from objectives to information; in forward affixing we work forward from information to objectives (S. M. Bridges *et al.*, 2000).

The three fundamental segments are given below:

(i)   Human interface.
(ii) Decision-making surmising motor.
(iii) Pools of data (putting away the information and Fuzzy principles).

Digital master can connect with the help of Knowledge-based System interface so as to ask and peruse the guidance from the new system. The derivation motor comprises the digital information Threats, digital fearmonger profiles, and digital assault methods.

10.3.4   Level 4: Rule-Based Model

In this model, overall engineering for knowledge-based system and the segments of a Fuzzy guideline-based derivation framework. The principle object of a Fuzzy standard-based framework are fuzzification – or fuzzifier module – Fuzzy guidelines, induction motor and defuzzifier.

Level 1. Fuzzification model: In the fuzzification, it contribution of the space of the information object area to calculate by set (Fuzzy). Developing a Fuzzy rationale participation capacities assume a vital job for Fuzzy principle-based structure. Three-sided participation work was utilized in numerous Fuzzy rationale-based applications (J.T. Yao *et al.*, 2005). In this examination three-sided participation capacities have been utilized in Figure 10.8.

Level 2. Characterizing Fuzzy guidelines: The Fuzzy standards comprise precursor and resulting as IF-THEN articulations. There are various principles, and they make a gathering which frames the reason for deduction (Singh Rathore, P *et al.*, 2020).

Level 3. Defuzzification: The coordination between the Fuzzy rationale control and the induction framework, by giving the fresh yield. Ordinary defuzzification techniques are centroid, bisector, mean estimation of greatest qualities, littlest estimation of most extreme qualities and biggest estimation of greatest. The transformation of a Fuzzy set to a solitary fresh worth is called defuzzification and opposite procedure is fuzzification.

There are number of most extreme worth can be anticipated of centroid strategies the gathering is kept up in the for the make a gathering the

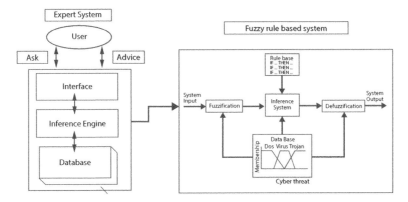

Figure 10.8 Model of rule-based system.

premise of obstruction the littlest and biggest worth is characterized in bisector instrument which is utilized in the model m of fuzzification of rule based, the fresh worth and Fuzzy worth is characterize on the numerical worth. The customary worth is needy of the interface to demonstrating the fresh output.

10.4 Digital Security

10.4.1 Cyber-Threats

Digital-based advancements are presently pervasive around the world. By far, most clients seek legitimate, proficient and individual goals. In any case, lawbreakers, psychological oppressors, and spies likewise depend vigorously on digital-based advances to help them reach their targets. These transgressors may get to digital-based advancements so as to refuse assistance, take or control information, or utilize a gadget to dispatch an assault against itself or another bit of gear. Elements utilizing digital-based advancements for illicit purposes take numerous structures.

10.4.2 Cyber Fault

System fault, digital fault, e-fault, electronic fault for the most part including crime where a system is the source, device, target, or spot of a fault. In this fault, the grouping are not selective and numerous exercises can be described as falling into at least one class. Moreover, in spite of the fact that the terms system fault or cybercrime are all the more appropriately limited to depicting crime where the system or system is an important piece of the fault, these terms are additionally some of the time used to incorporate conventional violations, for example, misrepresentation, burglary, coercion, falsification, and theft, in which systems or systems are utilized to encourage unlawful movement (J. Luo *et al.*, 2000). Digital fault is additionally a significant issue nowadays globally; the same number of individuals are hacking into the system frameworks.

System fault can comprehensively be characterized as crime including a data innovation framework, including illicit access (unapproved get to), unlawful block attempt (by specialized methods for non-open transmissions of system information to, from or inside a system framework), information obstruction (unapproved harming, cancellation, disintegration, modification or concealment of system information), frameworks impedance (meddling with the working of a system framework by contributing, sending, harming, erasing, falling apart, modifying or smothering PC information), abuse of gadgets, imitation (ID robbery), and electronic extortion. Below are the kinds of system faults:

    i.   Computing virus
   ii.   Fraud activity
  iii.   Malicious Code
  iv.   Denial-of-Service Attack
   v.   Hacking
  vi.   Crime (cyber)
 vii.   Cyber Terrorism
viii.   Information Warfare
  ix.   Cyber Stalking
   x.   Fraud and Identity Theft
  xi.   Crime (virtual)

10.4.3 Different Types of Security Services

Security mechanism is a significant component engaged with a system. At the point when clients share assets and information on a system, they ought to have the option to control who can get to the information or asset and what the client can do with it. A case of this is a document indicating the financial records of an organization. In the event that this document is on a record worker, it is imperative to have the option to control who approaches the document. Above and beyond, who can peruse and change the record, likewise, is a pivotal thought. This equivalent model additionally applies to a common printer. You should determine who can utilize the costly shading laser printer or, all the more explicitly, when an individual can utilize this printer. As should be obvious, security is a significant help on a system. System directors invest a lot of energy learning and setting up security. Security benefits frequently manage a client account database or something like the previously mentioned index administrations. The client's database frequently contains a rundown of username and random number. An individual needs to get to the system; he should sign on to the system. Signing on is like attempting to enter a place of business with a safety officer at the front entryway. Before you can enter the structure, you should check who you are against a rundown of individuals who are permitted to enter. Security benefits regularly are blended with different administrations. A few administrations added to a system can use the security administrations of the framework onto which they have been introduced.

10.5   Improvement of Cyber Security System (Advance)

At this point, current writing on digital security framework has been summed up, and the basic restrictions of the past featured. The structuring level incorporates characterizing digital security framework factors, information assortment for digital Threats, framework plan and usage. The levels are defined below:

10.5.1   Structure

The starting level in the new system is the foundation of information and yield factors. This errand is typically done by examining the difficult space. There is an interminable number of potential up-and-comers which ought to be limited to positive numbers which can be describe by Figure 10.9.

Input Variables	Abbreviation	Output Variables	Abbreviation
Cyber Techniques	**CT**	Software	**S**
Aim of Cyber Intruders	**ACI**	Hardware	**H**
Cyber Intruder's Target	**CIT**	User	**U**
Cyber Intruders	**CI**		

Figure 10.9 Input output variables.

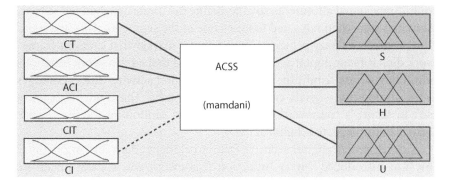

Figure 10.10 The structure of cyber security system.

10.5.2   Cyber Terrorism for Information/Data Collection

In this section, the development structure, the information on the human master. It defines clarifications like the human expert. Model can design typical analysis by the user. The information utilized for this work has been removed from a progression of surveys gathered from digital framework executives described in Figure 10.10. The important points are:

- DoS assaults, infection, rationale bomb, social designing.
- Out of administration, holding onto site page, assaults for dissenting, hold onto basic frameworks, catch secret data, framework control.

This investigation assesses digital fear-based oppressors who may assault correspondence frameworks, budgetary focuses, power plants, crisis administrations, transportation, water supplies, oil and petroleum gas appropriation stations. Individuals equipped for digital psychological oppression, for example, devoted unique staff, programmers and digital model system.

10.6   Conclusions

A specialist framework for digital security is dependent on Fuzzy principle. After meeting with digital specialists and framework directors, the data sources and yield of the framework were resolved. The deduction of the Fuzzy standards was completed utilizing the 'min' and 'max' administrators for Fuzzy convergence and association. Information space was isolated into multidimensional allotments so as to define the underlying guideline base. Activities were then relegated to every one of the segments. This investigation proposes a Fuzzy principle–based digital marker that cautions framework directors for digital activity. It discovered that a framework functions admirably whose condition matched with digital activity. The encourages some admonition signals created fair and square. The model's objective isn't to secure a framework anyway it targets cautioning the framework chairman for expected digital activity.

A development digital security framework dependent on Fuzzy guideline was introduced. Fuzzy induction framework was chosen for create cyber security system. The deduction of the Fuzzy standards was completed utilizing the 'minimum' and 'maximum' administrators for Fuzzy crossing point and association. Info space was separated into multidimensional parcels so as to plan the underlying principle base. Activities were then doled out to every one of the segments. Made sure about information correspondence over web and some other system is consistently under Threats of interruptions and abuses. So Intrusion Detection Systems have become a needful segment as far as PC and system security are concerned. There are different methodologies being used in interruption location; however, none of the frameworks so far are totally perfect. Along these lines, the mission of improvement proceeds. Fuzzy rationale strategy gives an approach to describe the loosely characterized factors, characterize connections between factors dependent on master human information and use them to process results. Fuzzy Expert System applied to data security field is adequate method for copying pro's dynamic capacity.

References

R. Chandia, J. Gonzalez, T. Kilpatrick, M. Papa, S. Shenoi, "Security strategies for SCADA networks," in: *Proceeding of the First Annual IFIP Working Group 11.10 International Conference on Critical Infrastructure Protection, Dartmouth College, Hanover, New Hampshire, USA, Mar. 19–21, 2007.*

Singh Rathore, P., Kumar, A., & Gracia-Diaz, V. (2020). A Holistic Methodology for Improved RFID Network Lifetime by Advanced Cluster Head Selection using Dragonfly Algorithm. *International Journal of Interactive Multimedia and Artificial Intelligence*, 6 (Regular Issue), 8. http://doi.org/10.9781/ijimai.2020.05.003

N. Bhargava, S. Dayma, A. Kumar and P. Singh, "An approach for classification using simple CART algorithm in WEKA," *2017 11th International Conference on Intelligent Systems and Control (ISCO)*, Coimbatore, 2017, pp. 212–216. doi: 10.1109/ISCO.2017.7855983

Naveen Kumar, Prakarti Triwedi, Pramod Singh Rathore, "An Adaptive Approach for image adaptive watermarking using Elliptical curve cryptography (ECC)", *First International Conference on Information Technology and Knowledge Management*, pp. 89–92, ISSN 2300-5963 ACSIS, Vol. 14 DOI: 10.15439/2018KM19

Neeraj Bhargava, Abhishek Kumar, Pramod Singh, Manju Payal, "An Adaptive Analysis of Different Methodology for Face Recognition Algorithm", *International Journal on Future Revolution in Computer Science & Communication Engineering (IJFRSCE)*, December 17, Volume 3 Issue 12, pp. 209–212.

N. Fovino, M. Masera, "Through the description of attacks: a multidimensional view", in: *Proceeding of the 25th International Conference on Computer Safety, Reliability and Security, Gdansk, Poland, Sep. 26–29, 2006.*

R. Shanmugavadivu, "Network Intrusion Detection System Using Fuzzy Logic", *Indian Journal of Computer Science and Engineering (IJCSE)*, vol. 2, 1, pp. 101–111, 2011.

S. M. Bridges, and R. B.Vaughn, "Fuzzy Data Mining and Genetic Algorithms Applied to Intrusion Detection", In *Proceedings of the National Information Systems Security Conference (NISSC), Baltimore, MD, 2000*, pp. 16–19.

Bharat Singh, Ravinder Singh and Pramod Singh Rathore. Article: Randomized Virtual Scanning Technique for Road Network. *International Journal of Computer Applications* 77(16):1–4, September 2013.

J.T. Yao, S.L. Zhao, and L.V. Saxton, "A Study On Fuzzy Intrusion Detection", In *Proceedings of the Data Mining, Intrusion Detection, Information Assurance, and Data Networks Security, SPIE*, Vol. 5812, Orlando, Florida, USA, 2005, pp. 23–30.

S. Mukkamala, G. Janoski, A. Sung, "Intrusion detection: support vector machines and neural networks." In: *Proceedings of the IEEE International Joint Conference on Neural Networks (ANNIE)*, St. Louis, MO, 2002, pp. 1702–1707.

Y. Yu, and H. Hao, "An Ensemble Approach to Intrusion Detection Based on Improved Multi-Objective Genetic Algorithm", *Journal of Software*, Vol. 18, No. 6, pp. 1369–1378, June 2007.

P. S. Rathore, A. Chaudhary and B. Singh, "Route planning via facilities in time dependent network," *2013 IEEE Conference on Information & Communication Technologies*, Thuckalay, Tamil Nadu, India, 2013, pp. 652–655. doi: 10.1109/CICT.2013.6558175

J. Cannady, "Artificial Neural Networks for Misuse Detection", in *Proceedings of the '98 National Information System Security Conference (NISSC'98)*, 1998, pp. 443–456.

W. Lee, S. Stolfo, and K. Mok, "A Data Mining Framework for Building Intrusion Detection Model", In *Proceedings of the IEEE Symposium on Security and Privacy*, Oakland, CA, 1999, pp. 120–132.

Pramod Singh Rathore, "An adaptive method for Edge Preserving Denoising, International Conference on Communication and Electronics Systems,

Institute of Electrical and Electronics Engineers & PPG Institute of Technology (2017). *Proceedings of the 2nd International Conference on Communication and Electronics Systems (ICCES 2017)*: 19-20 October, 2017.

J. Luo, and S. M. Bridges, "Mining fuzzy association rules and fuzzy frequency episodes for intrusion detection", *International Journal of Intelligent Systems*, Vol. 15, No. 8, pp. 687–704, 2000.

Neeraj Bhargava, Pramod Singh, Abhishek Kumar, Taruna Sharma, Priya Meena, "An Adaptive Approach for Eigenfaces-based Facial Recognition", *International Journal on Future Revolution in Computer Science & Communication Engineering (IJFRSCE)*, December 17, Volume 3, Issue 12, pp. 213–216.

A.N. Toosi, M. Kahani, "A new approach to intrusion detection based on an evolutionary soft computing model using neuro-fuzzy classifiers. *Computer Communications*, vol. 30, pp. 2201–221, 2007.

A. Tajbakhsh, M. Rahmati, A. Mirzaei, "Intrusion detection using fuzzy association rules", *Applied Soft Computing*, Vol: 9, No: 2, pp. 462–469, 2009.

B. Shanmugam, N. B. Idris, "Improved Intrusion Detection System Using Fuzzy Logic for Detecting Anamoly and Misuse Type of Attacks", in *Proceedings of the International Conference of Soft Computing and Pattern Recognition, 2009*, pp. 212–217.

O. Cordon, F. Gomide, F. Herrera, F. Hoffmann, L. Magdalena, "Ten years of genetic fuzzy systems: current framework and new trends", *Fuzzy Sets and Systems*, vol.141, no.1, pp. 5–31, 2004.

L.A. Zadeh, "Fuzzy sets", *Information Control*, vol.8, pp. 338–353,1965.

E.H. Mamdani, and S. Assilian, "An experiment in linguistic synthesis with a fuzzy logic controller", *Int. J. Man-Mach. Stud.*, vol.7, pp. 1–13, 1975.

J. Lu, G. Zhang, D. Ruan, *Multi-Objective Group Decision Making: Methods, Software and Applications with Fuzzy Set Techniques*, Imperial College Press, London, 2007.

J.C. Giarratano and G. Riley, "Expert systems principles and programming", MA, USA: PWS-KENT Publishing Company, 1989.

N.J. Nilsson, *Principles of Artificial Intelligence*, Palo Alto, CA. Tioga, 1980.

M. Schneider, G. Langholz, A. Kandel, and G. Chew, *Fuzzy Expert System Tools*, John Wiley & Sons, USA, 1996.

Hinson, G. 2008. Frequently Avoided Questions about IT Auditing - http://www.isect.com/html/ca_faq.html

Val Thiagarajan, B.E. 2002. BS 7799 Audit Checklist. - www.sans.org/score/checklists/ISO_17799_checklist.pdf

ISO IEC 27002 2005 Information Security Audit Tool - http://www.praxiom.com/iso-17799-audit.htm

Stepanova, D., Parkin, S. and Moorsel, A. 2009. A Knowledge Base for Justified Information Security Decision-Making. In *4th International Conference on Software and Data Technologies (ICSOFT 2009)*, 326–311.

Giarratano, J., and Riley, G. eds. 2002. *Expert Systems: Principles and Programming*. Reading, MA: PWS Publishing Company.

Tsudik, G. and Summers, R. 1990. *AudES - an Expert System for Security Auditing*. IBM Los Angeles Scientific Center.

11

Threat Analysis Using Data Mining Technique

Riddhi Panchal[1]* and Binod Kumar[2]

[1]*Research Scholar (Computer Management), Savitribai Phule Pune University, Maharashtra*
[2]*Dr. Professor, JSPM'S Rajarshi Shahu College of Engineering, Pune, Maharashtra*

Abstract

With the advancement in information technology, internet usage plays a vital role in daily activities, and due to this cyber terrorism has been increasing rapidly. Cyber criminals commit numerous cyber attacks such as Phishing, Denial of service, Password attack, etc. Due to lack of computation methods, existing technical approaches are not sufficient to investigate and control cyber attacks. Therefore the current scenario requires a more advanced approach to fix cyber-attack issues. The goal of this chapter is to analyze cyber threats and to demonstrate how artificial intelligence and data mining approaches can be effective to fix cyber attack issues. The field of artificial intelligence has been playing an increasingly vital role in analysing cyber threats and improving cyber security as well as safety. Mainly three aspects are discussed in this chapter. First, the process of cyber-attack detection which will help to analyse and classify cyber incidents. Second, forecasting upcoming cyber attacks and controlling cyber terrorism. Finally, the chapter focuses on the theoretical background and practical usability of artificial intelligence with data mining approaches for addressing the above issues through detection and prediction.

Keywords: Cyber attack, cyber security, cyber criminals, detection, prediction, classification

**Corresponding author*: rids.panchal@gmail.com

Neeraj Bhargava, Ritu Bhargava, Pramod Singh Rathore, and Rashmi Agrawal (eds.) Artificial Intelligence and Data Mining Approaches in Security Frameworks, (197–208) © 2021 Scrivener Publishing LLC

11.1   Introduction

In recent years, the whole world has become dependent on ICT information and communication technology for professional work, entertainment, education, and social life. The reason behind this dependency is the popularity of IoT (Internet of Things), marvelous computer network expansion, and the large amount of applications used by various groups as well as individuals for their personal and professional use. The cyber security demand has increased due to different cyber attacks like unauthorized access, denial-of-service attack, computer malware, etc. The computer and network security system results in cyber security.

Encryption and firewall type of systems are available to manage cyber attacks. Presently, conventional solutions such as firewalls are not able to carry out their jobs. Due to lack of computation methods, existing technical approaches are not sufficient to investigate and control cyber attacks. Therefore the current scenario requires a more advanced approach to fix the above cyber-attack issues.

This chapter presents the machine learning (ML) and data mining (DM) methods for cyber security. The goal of the current chapter is to analyze cyber threats and to demonstrate how artificial intelligence and data mining approaches can be effective to fix cyber-attack issues.

The field of artificial intelligence has been playing an increasingly vital role in analysing cyber threats and improving cyber security as well as safety. Nowadays Artificial Intelligence is a world-changing tool, which enhances human abilities in a variety of areas (A. Meena et al., 2018).

Mainly three aspects are discussed in this chapter. First, the process of cyber-attack detection, which will help to analyze and classify cyber incidents. Second, forecasting upcoming cyberattacks and controlling cyber terrorism. Finally, the chapter focuses on theoretical background and practical usability of artificial intelligence with data mining approaches for addressing the above issues through detection and prediction (A. Okutan et al., 2018; Bharat Singh et al., 2013).

After the section 11.1 introduction, the remaining sections are structured in the following manner. Section 11.2 provides the background and related work of cyber attack and cybersecurity with data mining techniques (Artificial intelligence and machine learning).

In Section 11.3, various Data Mining–based methods are discussed for the detection of cyber ttacks. Section 11.4 describes the process of cyber-attack detection based on data mining.

Finally, the conclusion of the chapter is in Section 11.5, which also highlights future scope.

11.2   Related Work

Various existing research works are analyzed to understand the purpose of the current chapter.

Cyber Security
It is also called as IT information technological security. Cyber security is one type of technology body and process that protect computer networks, software and computer devices from different types of damage (Ch. Rupa *et al.*, 2020). Nowadays cyber security is in demand for different government sectors, Defence and army, financial institutions, medical and health organisations. The aim is to sustain the confidentiality, veracity, as well as availability of information management systems by a cyber defence system. Research scholars from academia, government sectors and private industries has been engaged in designing and executing different types of cyber defence systems (as described in Figure 11.1) (Coombs T., 2018, E. Doynikova *et al.*, 2020).

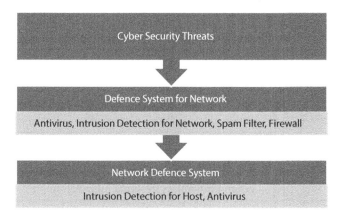

Figure 11.1 Conventional system for cyber security.

Landscape of Cyber Threat
Although the common threat categories have not changed, the landscape of adversaries and troublemakers in computer security has evolved over time. For that purpose it is essential to have good knowledge of the variety of attacks possible (Chio C *et al.*, 2018; H. Tianfield *et al.*, 2017).

Common cyber attacks are listed in Table 11.1.

Table 11.1 Common cyber attacks and their description.

Cyber attacks	Description
Malware/Trojan virus	Software purposely designed to cause damage to the computer system
Spyware	Malware installed on a computer system without permission for the purposes of infiltration and information collection
Adware	Advertising supported malware
Rootkit	Causing havoc in the computer background
Backdoor	Planned hole placed allowing future access without perimeter protection.
Bot	Self-propagating malware
Botnet:	Huge network of bots.
Exploit	Code that benefits from software vulnerability
Scanning	By sending request to computer system in a brute-force manner for searching weak point, information gathering and vulnerability.
Sniffing	Spying virus which allow to record personal information
Keylogger	Record key pressed information with piece of software/hardware
Spam	Spread from one system to another system by means of email
ATO	Account takeover; Gaining access to another's account
Phishing	Attackers show themselves as legal business to fool and get an individual's personal information
DoS	Denial of service attacks through high-volume bombardment and/or malformed requests

Artificial Intelligence (AI) and Machine Learning (ML)

AI is the attempt by individuals to make machines smarter. AI definition is slightly more controversial than machine learning definition. Artificial Intelligence is defined as machine-oriented decision to accomplish human-level brain tasks, and machine learning is defined as learning from past

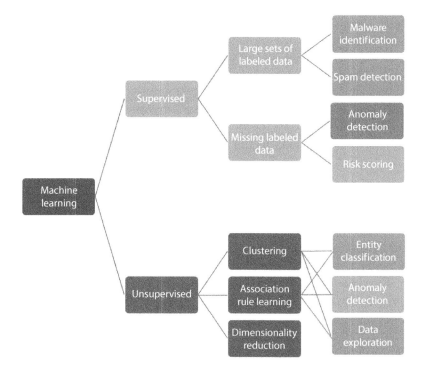

Figure 11.2 Cyber security with supervised and unsupervised machine learning.

data to predict the future (Maochao Xu *et al.*, 2018; K.P. Barabde *et al.*, 2019). There are two main types of machine learning:

1. Supervised machine learning: Labeled data provides direct feedback and predicts future.
2. Unsupervised machine learning: No labels, no feedback, only finds hidden structure in data.

Figure 11.2 explains cyber security with supervised and unsupervised machine learning.

11.3   Data Mining Methods in Favor of Cyber-Attack Detection

The current section depicts the Data Mining methods in favour of cyber attack and related to its security. Data mining techniques like association rules, classification, clustering are generally used to detect different types of cyber attack.

Classification

Classification, also called classifier, which assigns data objects to predefined classes. First of all, by analyzing one training set, classifier is trained. Here training set is invented from data instance and its associated class label. This type of approach is supervised learning as training instance's class label is made available. Secondly, by using training classifier forecasting process of class for the unlabelled data instance is done. All the classes are predetermined in the training phase. Generally classification has two types of cases (N. Bhargava *et al.*, 2017).

A. Binary Classification: Here only two classes are engaged.
B. Multiclass classification: Here several classes are engaged.

For misuse as well as anomaly detection, classification may be used. In audit data, every data instance is labeled as Normal data or Abnormal data. The classification algorithm is applied on audit data for the purpose of training the classifier, which will be used for the prediction purpose whether the new instance data will be "normal" or "abnormal". A few popular classification methods are K-nearest neighbor, Naive Bayes classifier, decision tree, support vector machine, fuzzy logic, Artificial Neural Network (Nan Sun *et al.*, 2019).

Decision Trees

It is the tree type of formation with leaves. The decision tree symbolizes classification where its branches symbolize union of features that will lead to classification. "IF THEN" rule is the pillar of a decision tree. The simplest as well as interpretable structure permit decision tree for solving problems like multi-type attribute. Decision trees are used to manage noise data or missing values [5]. Decision trees provides very simple and easy implementation.

K-Nearest Neighbor (KNN)

This algorithm is basically classification algorithm, which is very easy. All the present cases that are training data, accumulated and upcoming new cases that are test data, will be classified depending on its similarity measure of given feature space. The new instance data distance from new case and existing cases is computed then test data is selected to the classes which will more familiar amongst its KNN (K-nearest neighbor).

Here, if value of k=1,

Then it is allocated to its nearest neighbor class.

If the k value is large then more prediction time is required (Naveen Kumar *et al.*, 2019; Neeraj Bhargava *et al.*, 2017).

Association Rule Mining
It will find the relation among different variables present in database. After data integration and data cleaning association rule mining will be applied. In association rule mining, first phase for request item set generation then next stage rule generation, which will create cyber attack attribution rules (Singh Rathore *et al.*, 2020).

Suppose IF (X AND Y) THEN Z.

Means if X and Y both will available then z is also available. For cyber-attack detection the same association rule is applied. If (Proof1 AND Proof2) THEN criminal identification.

Figure 11.3 describes how to create cyber-attack attribution rules (P. S. Raj *et al.*, 2017; P. S. Rathore *et al.*, 2013).

Clustering
The term clustering means distributing based on some similarity. In the intrusion detection clustering learns from audit data without system administrator by providing details of different attack classes (shown in Figure 11.4) [4]. Clustering is categorised into two types, namely,

A. Hard Clustering: item can be assigned into one cluster only.
B. Soft Clustering: item can be assigned into multiple cluster.

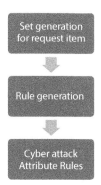

Figure 11.3 Cyber-attack attribution rules through association rule mining.

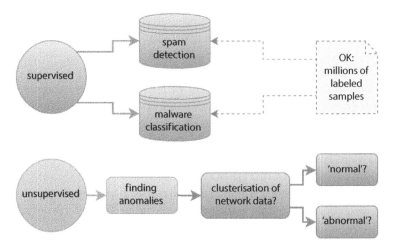

Figure 11.4 Supervised learning (Classification) and Unsupervised learning (Clustering).

11.4   Process of Cyber-Attack Detection Based on Data Mining

A detailed description of the process of Cyber-Attack Detection based on Data Mining is shown in Figure 11.5. In stage 1 Data Processing includes system monitoring as well as data capturing through different sensors, system or network logging as well as sniffing agents or daemons (Pramod Singh Rathore *et al.*, 2017).

Data mining tools for cyber security
Data mining requires a large data set and different data mining algorithms are applied to that large data set. This process also needs some statistical analysis, which is not possible manually. Presently many data mining tools are available; some tools are free. Popular data mining tools are listed in Table 11.2.

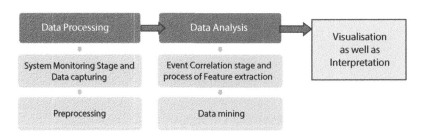

Figure 11.5 Different stages for detecting cyber attack through data mining.

Table 11.2 Popular Data Mining Tools for cyber security.

Sr. no.	Data Mining Tool
1	SAS Data mining Statistical Analysis System
2	Teradata
3	R-Programming
4	RapidMiner
5	Oracle BI Business Intelligence
6	KNIME
7	Tanagra
8	Weka Waikato Environment for Knowledge Analysis
9	Python
10	IBM SPSS Modeler

11.5   Conclusion

Data mining has enough strength for malware detection. It enables the dissection of a large amount of data and concentrates new learning from it. It has the ability to discover known as well as zero-day attacks. This chapter is about threat analysis using different data mining methods, algorithms and tools. The most significant data mining stage in cyber threat analysis is dataset for training and testing. In the future, more advanced deep learning Algorithms will be analyzed for cyber-attack detection and prediction.

References

A. Meena, "Data Mining Techniques Used In Cyber Security", *International Journal on Future Revolution In Computer Science & Communication Engineering* Issn: 2454-4248 Volume: 4 Issue: 11 19–21, Nov. 2018.

A. Okutan *et al.*, "Forecasting Cyberattacks with Incomplete, Imbalanced, and Insignificant Data", *Cybersecurity*, 2018.

Bharat Singh, Ravinder Singh and Pramod Singh Rathore. Randomized Virtual Scanning Technique for Road Network. *International Journal of Computer Applications* 77(16):1–4, September 2013.

Ch. Rupa *et al.*, "Computational System to Classify Cyber Crime Offenses Using Machine Learning", Www.Mdpi.Com/Journal/Sustainability 2020, 12, 4087; Doi:10.3390/Su12104087

Chio C. & Freeman D. (2018), "Machine Learning and Security: Protecting Systems with Data and Algorithms", United States of America, O'Reilly Media, Inc., 2018.

E. Doynikova *et al.*, "Attacker Behaviour Forecasting Using Methods of Intelligent Data Analysis: A Comparative Review and Prospects", Www.Mdpi.Com/Journal/Information 2020, 11, 168; Doi:10.3390/Info11030168

H. Tianfield," Data Mining Based Cyber-Attack Detection", *System Simulation Technology* (Issn1673-1964), Vol. 13, No. 2, Apr. 2017, pp. 90–104.

Iqbal H. Sarker *et al.*, " IntruDTree: A Machine Learning Based Cyber Security Intrusion Detection Model, *Symmetry* 2020, 12, 754; doi:10.3390/sym 12050754

K.P.Barabde, V. Y. Gaud, "A Survey Of Data Mining Techniques for Cyber Security", *Journal of Emerging Technologies and Innovative Research (JETIR)*, May 2019, Volume 6, Issue 5.

Maochao Xu, "Modeling And Predicting Cyber Hacking Breaches", *IEEE Transactions on Information Forensics And Security*, Vol. 13, No. 11, November 2018.

N. Bhargava, S. Dayma, A. Kumar and P. Singh, "An approach for classification using simple CART algorithm in WEKA," *2017 11th International Conference on Intelligent Systems and Control (ISCO)*, Coimbatore, 2017, pp. 212–216. doi: 10.1109/ISCO.2017.7855983

Nan Sun *et al.*, "Data-Driven Cybersecurity Incident Prediction: A Survey", *IEEE Communications Surveys & Tutorials*, Vol. 21, No. 2, Second Quarter 2019

Naveen Kumar, Prakarti Triwedi, Pramod Singh Rathore, "An Adaptive Approach for image adaptive watermarking using Elliptical curve cryptography (ECC)", *First International Conference on Information Technology and Knowledge Management*, pp. 89–92, ISSN 2300-5963 ACSIS, Vol. 14 DOI: 10.15439/2018KM19

Neeraj Bhargava, Abhishek Kumar, Pramod Singh, Manju Payal, "An Adaptive Analysis of Different Methodology for Face Recognition Algorithm", *International Journal on Future Revolution in Computer Science & Communication Engineering (IJFRSCE)*, Volume 3 Issue 12, December 17, 2017, pp. 209–212.

P. S. Raj, "Role of Data Mining In Cyber Security", *International Journal of Engineering Science and Computing*, July 2017

P. S. Rathore, A. Chaudhary and B. Singh, "Route planning via facilities in time dependent network," *2013 IEEE Conference on Information & Communication Technologies*, Thuckalay, Tamil Nadu, India, 2013, pp. 652–655. doi: 10.1109/CICT.2013.6558175

Pramod Singh Rathore, "An adaptive method for Edge Preserving Denoising." International Conference on Communication and Electronics Systems,

Institute of Electrical and Electronics Engineers & PPG Institute of Technology (2017). *Proceedings of the 2nd International Conference on Communication and Electronics Systems (ICCES 2017)*: 19–20, October 2017.

Singh Rathore, P., Kumar, A., & Gracia-Diaz, V. (2020). A Holistic Methodology for Improved RFID Network Lifetime by Advanced Cluster Head Selection using Dragonfly Algorithm. *International Journal of Interactive Multimedia and Artificial Intelligence,* 6 (Regular Issue), 8. http://doi.org/10.9781/ijimai.2020.05.003

12

Intrusion Detection Using Data Mining

Astha Parihar[1]* and Pramod Singh Rathore[2]

[1]M.Tech Scholar, MDS University, Ajmer, India
[2]Aryabhatta Engineering College, Ajmer, India

Abstract

Presently, the internet is an almost universal method of communication for both individuals and businesses. Due to increased use of the internet, its security perspective is becoming more important every day for numerous network intrusion detection systems (IDS) from several attacks. Several IDS are located at heterogeneous locations of networks to preserve it. Various methods are used for detection of attacks or fraud and can be enforced in decision tree perspective. It gives the simplest way to recognise the maximum right area to select, manage and form optimize decision regarding their identification from greatest set of data. This chapter explores the modern intrusion detection with a distinctive determination perspective of data mining. This discussion focuses on major facets of intrusion detection strategy, that is, misuse detection. It focuses on identifying attacks, information or data which is present on the network using C4.5 algorithm, which is a type of decision tree technique and also it helps to enhance the IDS system to recognize types of attacks in network. For this attack detection, KDD-99 dataset is used; it contains several features and a different class of general and attack-type data.

Keywords: Intrusion detection, data mining, web intrusion detection system (WIDS), decision tree, C4.5 decision tree, knowledge discovery dataset-99 (KDD-99)

12.1 Introduction

In recent times, every enterprise and institution uses the internet for verbal exchange as well as, for commercial enterprise media, for reaching the

---
*Corresponding author: asthaparihar9@gmail.com

Neeraj Bhargava, Ritu Bhargava, Pramod Singh Rathore, and Rashmi Agrawal (eds.) *Artificial Intelligence and Data Mining Approaches in Security Frameworks*, (209–228) © 2021 Scrivener Publishing LLC

patron. Due to the fact the use of the internet accelerated, the increase of community fraud also increased, for which low self-guarantee in community structures connectivity and their assets has expanded the potential harm due to the presence of fraudulent activity, which is launched towards the systems from a long way off property. It's exceptionally difficult to save everyone from extortion by methods for utilizing firewalls on the grounds that on each event explicit misrepresentation consolidates obscure shortcomings or bugs. Therefore, continuous interruption recognition frameworks are utilized to hit upon fraud and broadly used to stop an attack being developed; it offers an alert hint to the approved client or network director of the presence of malignant interest or the presence of frauds.

The aim of IDS is to discover interruptions directly into a PC or organization, by noticing different organization sports or characteristics. Here interruption alludes to any arrangement of developments that compromises the honesty, accessibility, or classification of an organization's helpful asset.

Intrusion detection consists of a ramification of device and strategies which incorporate device mastering, records, statistics mining, and so forth for the identification of an assault. In contemporary years, information mining method for community intrusion detection device has been giving immoderate accuracy and right detection of several types of fraud. The choice tree technique is one of the intuitionist and frank classification strategies in truth mining which may be used for this purpose. It has a high-quality benefit in extracting features and policies. So, the choice tree gives more significance to intrusion detection. The tree is built with the resource of identifying attributes and their related values in an awesome way for users to test the input statistics at each middleman node of the tree. After the tree is used, it is able to recommend newly coming records with the aid of manner of traversing, preliminary from a root node to the leaf node by using traveling all the internal nodes within the direction relying upon the check environment of the attributes at every node. The number one hassle in building a choice tree is which price is chosen for splitting the node of the tree (A. S. Georghiades *et al.*, 2001).

12.2   Essential Concept

In this segment, we discussed IDS with its class strategies and about numerous attack classes. It also makes a speciality of how the selection tree is built and various strategies within the decision tree for creating a right choice.

12.2.1   Intrusion Detection System

An intrusion detection system (IDS) is a machine that displays units network traffic for suspicious interest and problems alert while such interest is determined. It's for a software utility that scans a network or a system for a harmful activity or coverage breaching. Notification of any malicious assignment or violation is normally sent both to an administrator or accumulated centrally using security facts and event control (SIEM) devices. A SIEM device integrates outputs from more than one source and makes use of alarm filtering techniques to distinguish malicious interest from fake alarms.

Despite the fact that intrusion detection structures reveal networks for potentially malicious activity, they are also disposed to fake alarms. As a result, companies want to fast-track their IDS products once they first deploy them, so they can find out what regular site visitors at the community look like as compared to the malicious lobby (Bharat Singh *et al.*, 2013).

In spite of the way that interruption recognition structures uncover networks for possibly malignant action, they're additionally arranged to counterfeit alerts. Thus, organizations need to track their IDS items once they initially send them. The interruption discovery structures will catch what normal site guests at the network look like when contrasted with malignant activity (C. Liu and H. Wechsler, 2000).

In accordance with the position method of IDS, it tends to be sorted as host-based and network-based absolutely frameworks. In host-based IDS, its miles blessing on each host that wants to follow. It can decide whether an attempted attack is a triumph and can run over neighborhood fraud. In organization-based absolutely frameworks are observed the organization guests from unapproved get right of section to with the guide of which the hosts are making a comfortable association with have frameworks. This component takes less expense for arrangement, and it's additionally encouraging for sorting out fraud to and from more than one host (shown in Figure 12.1).

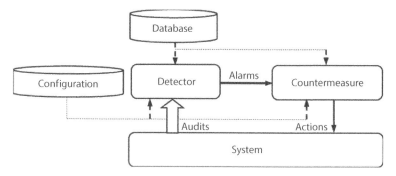

Figure 12.1 Intrusion detection system.

12.2.2   Categorization of IDS

IDSs can be categorized into many types based on the deployed platform to discover attacks and depending on the enter records that accumulated from exclusive sources such as system name, audit log, consumer or gadget activates, software procedure, and network visitors to evaluation and stumble on attack. Also IDS may be classified primarily based on the kind of attack that may be detected by means of each type (described in Table 12.1) (Frank Y. Shih *et al.*, 2008).

Table 12.1 IDS type comparative.

IDS	Base	Input data	Attacks that IDS detect
HIDS	Host	System Composition, activity of application, System logs, System order running process, access of files and modification	Key stroke logging, identity theft, unauthorized get entry to, spamming, malicious process, botnet hobby, adware-utilization.
WIDS	Network	Organization Traffic parcel, Prior occasions, client profile	TCP SYN fraud, divided bundle attack, Cross-Site Scripting (XSS), Cross-Site Request Forgery (CSRF)
CIDS	Web Server + Host	Normal usage of a protocol, http, structured question language (square) protocol, utility-level traffic, and instructions, audit records, information sources of running programs, and log fill	CANCEL DOS attack, BYE DOS attack, INVITE Request Flooding Attack, Media spamming, RTP packets flooding
Hybrid Based IDS	Host+Network	In line with systems hybrid	In line with systems hybrid

12.2.2.1   Web Intrusion Detection System (WIDS)

Web intrusion detection structures (WIDS) are installations at a deliberate factor within the network to observe traffic from all devices on the network. It performs a commentary of passing traffic on the entire subnet and matches the visitors this is passed at the subnets to the collection of acknowledged attacks. Once an attack is identified or peculiar behaviour is located, the alert can be despatched to the administrator. An example of a WIDS is putting it on the subnet in which firewalls are placed in an effort to see if a person is attempting to crack the firewall.

A WIDS is typically deployed or positioned at strategic factors for the duration of the community, meant to cover one's locations where site visitors are most likely to be liable to fraud. Generally, it's applied to complete subnets, and it tries to suit any site visitors passing by to a library of recognized attacks. It passively seems at community visitors coming through the points on the community on which it's deployed. They may be distinctly smooth and may be made difficult for intruders to hit upon. This indicates an interloper may not realize their capacity fraud is being detected with the aid of the WIDS.

Net-based totally intrusion detection machine analyses a huge quantity of network site visitors, which means they once in a while have low specificity. This indicates now and again they might miss a fraud or may not come across something taking place in encrypted traffic. In some instances, they might want extra manual involvement from an administrator to make certain they are configured correctly (shown in Figure 12.2) (G.R.S. Murthy and R.S. Jadon, 2009; J. C. Caicedo and S. Lazebnik, 2015).

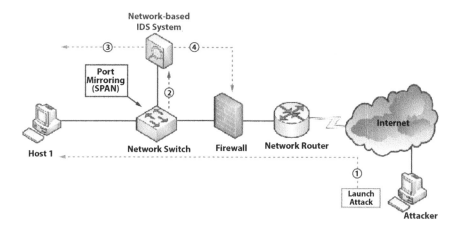

Figure 12.2 Web-based intrusion detection system.

12.2.2.2 Host Intrusion Detection System (HIDS)

Host intrusion detection systems (HIDS) run on impartial hosts or gadgets in the community. HIDS monitors the incoming and outgoing packets from the device handiest and will alert the administrator if a suspicious or malicious pastime is detected. It takes a photo of present machine files and compares them with the previous photo. If the analytical system documents have been edited or deleted, an alert is dispatched to the administrator to research. An instance of HIDS utilization can be seen on project-critical machines, which are not expected to exchange their layout.

The HIDS runs on all the gadgets in the organization with admittance to the web and different pieces of the endeavour organization. HIDS have a few focal points over NIDS, because of their capacity to look all the more carefully at inward traffic, just as filling in as a second line of protection against vindictive bundles a NIDS has neglected to identify.

It looks at the complete device's report set and compares it to its preceding "snapshots" of the record set. It then appears at whether or not there are giant differences outdoor ordinary enterprise use and alerts the administrator as to whether there are any lacking or extensively altered documents or settings. It mostly makes use of host-primarily based actions such as application use and files, report access throughout the gadget, and kernel logs (Jyh-Yeong Chang and Jia-Lin Chen, 2001).

Network and host-based intrusion detection systems are the maximum not unusual methods of expressing this category, and also you gained locate NIDS cited very frequently in this space. It can be thought of simply as a form of NIDS (shown in Figure 12.3).

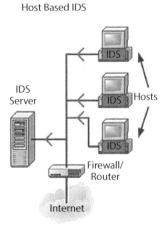

Figure 12.3 Host-based Intrusion Detection System

12.2.2.3 Custom-Based Intrusion Detection System (CIDS)

CIDS incorporates a device or agent that would always reside at the front stop of a server, controlling and deciphering the protocol between a person/ tool and the server. It is trying to ease the webserver via often tracking the HTTPS protocol flow into and take transport of the associated HTTP protocol. As HTTPS is unencrypted and earlier than proper away coming into its net presentation layer then this gadget could need to live in this interface, between to apply the HTTPS (N. Bhargava *et al.*, 2017).

12.2.2.4 Application Protocol-Based Intrusion Detection System (APIDS)

APIDS is a system or agent that normally lives within a set of servers. It identifies the intrusions with the aid of monitoring and interpreting the conversation on software-precise protocols. As an instance, this may monitor the sq. Protocol expresses to the middleware because it transacts with the database within the webserver (shown in Figure 12.4).

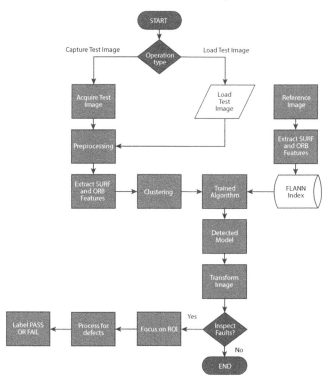

Figure 12.4 Application Protocol-based Intrusion Detection System (APIDS).

12.2.2.5   Hybrid Intrusion Detection System

A hybrid intrusion detection gadget is made with the aid of the combination of two or extra techniques of the intrusion detection gadget. In the hybrid intrusion detection device, host agent or system records are blended with community facts to broaden an entire view of the network device. A hybrid intrusion detection machine is extra effective in the assessment of the opposite intrusion detection device. Prelude is an example of hybrid IDS (shown in Figure 12.5) (Naveen Kumar *et al.*, 2018).

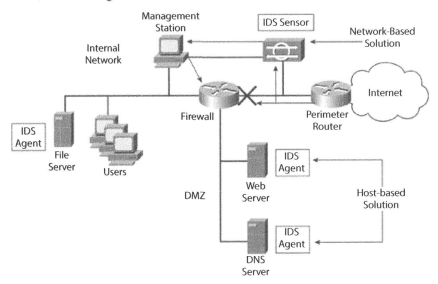

Figure 12.5 Hybrid intrusion detection system.

12.3   Detection Program

The current era of business intrusion detection structures in big element node-based totally and client detection. For example, present-day device absolutely lacks the capability to come across.

Lack of fraud detection in commercial structures, majorly look at hobby is to detection of variation, in location of mere extensions of misuse detection. Systems that contain two strategies are useful for study also. Evaluating the problem for variation detection is required to remove fake indication because every hobby outside a recognized profile will prompt an alarm. Sincerely, the wrong indication price is a finite problem in IDS.

Multiplied community velocity, swapped networks, the product program of encryption has brought roughly a design towards have based complete discovery. Another intriguing new methodology is transported interruption discovery, in which have based structures uncover some of the broadcast on the organization and move the checked knowledge to the main site page (P. S. Rathore *et al.*, 2013).

12.3.1 Misuse Detection

Typical, IDs generation is premature and hastily merging. Inside the modern domain, new suppliers appear habitually anyway are routinely ingested through others. On the examinations front, a dispersion of cycles is being explored. Yet, a standard hypothetical structure keeps on being deficient.

The significant methodologies that have been proposed for misuse discovery are master frameworks, signature investigation, state-progress examination, and information mining. Approaches have additionally been proposed including shaded Petri nets and case-based thinking.

Misuse detection searches for acknowledged styles of fraud. This is the strategy hired by means of the contemporary era of commercial intrusion detection systems. A disadvantage of this approach is that it is able to most effectively detect intrusions that comply with pre-described patterns (Pramod Singh Rathore, 2017; Neeraj Bhargava *et al.*, 2017).

The most important process which is introduced for anomaly detection is ES (Expert System).

12.3.1.1 *Expert System*

The ES deals with illegal location utilizing a fixed set of rules to depict misrepresentation. Review occasions are converted into records wearing their semantic significance inside the expert machine. An induction motor at that point draws in ends utilizing these rules and data.

Examples of such ESs are GIDS (Grid Intrusion Detection System), C-BEST (Construction Based Expert System), etc. (Neeraj Bhargava *et al.*, 2017).

IDX is an information-based model interruption recognition master framework for Unix System V. It consolidates information on the objective framework, history profiles of clients' past exercises, and interruption identification heuristics. The outcome is an information-based framework equipped for distinguishing explicit infringement that happen on the objective framework. A special component of GIDS is that it incorporates realities portraying the objective framework and heuristics typified in decides that identify specific infringement from the objective framework review trail. IDX is subsequently working framework subordinate.

C-BEST (created at SRI) is a standard-based, forward-binding master a framework that has been applied to signature-based interruption discovery for numerous years. The principle thought is to indicate the qualities of a vindictive conduct and afterward screen the flood of occasions produced by the framework movement, planning to perceive an interruption signature. C-BEST is a broadly useful procurable master framework coat, wearing a standard definition language that is sufficiently straightforward to be utilized by non-specialists. The framework was first conveyed in the MIDAS ID framework at the National Computer Security Centre (NCSC). Afterward, C-BEST was picked as the standard-based induction motor of NIDES, a replacement to the IDES model. The C-BEST master framework shell is additionally utilized in EMERALD's master, a nonexclusive mark investigation motor (shown in Figure 12.6).

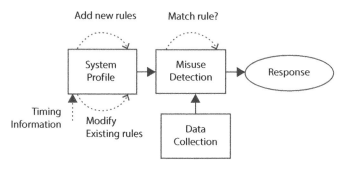

Figure 12.6 Misuse detection expert system (MDES).

12.3.1.2   Stamp Analysis

Mark assessment changes the semantic portrayal of extortion into records that might be found inside the review way in a fair manner. Instances of such insights envelop the successions of review events that extortion create or examples of information that can be looked for inside the review trail.

Frameworks that utilize signature investigation incorporate Haystack, NetRanger RealSecure and MiSig (Misuse Signatures).

- Stack is a misuse detection device that permits air force security officials to hit upon the misuse of Unisys mainframes. Working at the reduced audit trails records, it performs misuse detection based totally on behavioural constraints imposed through proper safety regulations and on fashions of traditional consumer behaviour.

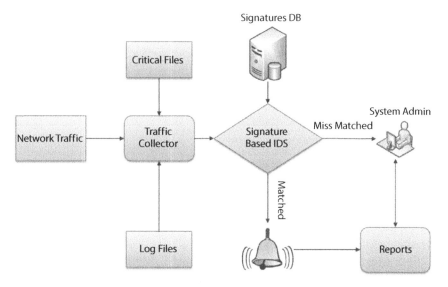

Figure 12.7 Signature-based analysis in IDS.

- Net-Ranger comprises two modules: sensors and executives. Sensors are network security video show units that inspect the organization traffic on an organization segment and the logging measurements created with the guide of cisco switches to stagger on organization essentially based assaults. Chiefs are liable for the control of a gathering of sensors and can be based progressively to control enormous organizations as shown in Figure 12.7.
- Real secure (progressed at web wellbeing structures) comprises three modules: network motors, machine advertisers, and heads. The organization motors are network video show units outfitted with misrepresentation marks that are coordinated against the guests on an organization hyperlink. The gadget dealers are have based interruption recognition frameworks that show security tricky log records on a number. Those modules report their finds to the essential administrator, which shows the records to the individual and presents functionalities for distant administration gadget dealers and network motors.
- Missing applies an excessive-stage language for abstract signatures. It attempts to triumph over certain limitations

of traditional misuse detection systems, which include the confined expressiveness of signatures expressed in low-level language and fixed tracking algorithms for misuse which have trouble adapting to a converting going for walks environment or safety goals. Via its excessive-level language, missing can constitute misuses in an easy form with high expressiveness.

12.3.1.3   Data Mining

Statistics (information) sorting prefers a way of non-trivial escarpment of inner, formerly unnamed, and doubtlessly useful information from databases. Models for intrusion detection), and automatic discovery of concise predictive pointers for intrusion detection.

IDS uses sorting (mining) encompass JAM. MARAD and automatic search pointer for IDS system. Instance anomaly detection system uses information sorting encompasses JAM and MARAD.

JAM makes use of facts mining techniques to find out patterns of intrusions. It then applies a meta-reading classifier to examine the signature of attacks. The association rules set of rules determines relationships among fields within the audit path data and the common episodes of a fixed of guidelines fashions sequential styles of audit activities. Features are then extracted from each algorithm and used to compute models of intrusion conduct. The classifiers assemble the signature of attacks. So essentially, records mining in JAM build a misuse detection model. JAM uses fact mining strategies to find out styles of intrusions. It then applies a meta-reading classifier to analyse the signature of attacks. The affiliation rules algorithm determines relationships between fields inside the audit trail statistics and the common episodes of a fixed of rules models sequential kinds of audit activities. Features are then extracted from each algorithm and used to compute fashions of intrusion conduct. The classifiers assemble the signature of fraud. So essentially, data mining in JAM builds a misuse detection model.

JAM originate classifiers the usage of a protocol gaining knowledge of application on schooling statistics of machine usage. The device is examined with information from ship e-mail-based totally fraud, and with network attacks the use of TCP dump statistics. MARAD identification makes use of fact mining to create policies for misuse detection. The inducement is that current structures require big manual effort to expand regulations for misuse detection.

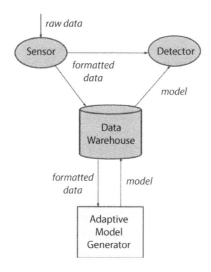

Figure 12.8 Data mining in intrusion detection system.

MARAD ID applies information mining to review information to figure models that precisely catch standards of conduct of interruptions and typical exercises (shown in Figure 12.8) (Naveen Kumar *et al.*, 2018).

12.4   Decision Tree

Statistics (information) mining is the technique for coming across information from massive set of data assist of information and synthetic intelligence techniques to solve complicated actual-life problems. The selection tree is a vital type of method within the facts mining class. A preference tree is described as a flowchart-like or tree-like shape from special verities of facts. Indecision tree, every internal node represents a check on a feature, while every stem represents the very last consequences of the test and each leaf node represents a class label. The route from the basis node to the leaf node represents the form of regulations. From an intrusion detection attitude, class algorithms can distinguish community statistics as fraud, benign, scanning, or some other category of interest using records like deliver/ destination ports, IP addresses, and the range of bytes dispatched in some unspecified time in the future of a connection (P. S. Rathore *et al.*, 2013).

An expansion tree classifier has an easy shape that is perfectly saved and labelled new information. Classifiers in decision tree contains many algorithm, such as CART, C4.5, and ID3.

12.4.1　Classification and Regression Tree (CART)

Decision Trees are normally used in data mining with the goal of creating a model that predicts the fee of a target (or established variable) based on the values of several inputs (or unbiased variables). In the modern-day post, we talk about the CART decision tree methodology. The CART or classification and regression tree technique was introduced in 1984 by Leo Breiman, Jerome Friedman, Richard Olsen, and Charles stone.

CART builds a binary tree that means a dataset which contains node most efficaciously can be splits in orporations. CART can take care of several shapes of data like every unique and arithmetic information. CART uses the Gini index for deciding on attributes. The function with the most crucial discount in impurity is used for splitting the nodes of the dataset. It uses charge-complexity pruning and additionally generates regression wood (N. Bhargava *et al.*, 2017).

Classification Tree: wherein the goal variable is specific and the tree is used to identify the "Class" within which a target variable could probably fall into (Figure 12.9).

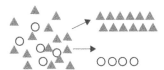

Figure 12.9　Classification tree.

Regression time: In which the target variable is non-stop and the tree is used to be expecting its value (Figure 12.10).

Figure 12.10　Regression tree.

12.4.2　Iterative Dichotomise 3 (ID3)

This set of guidelines was layout-primarily based mostly on the standards of Occam's razor, with the concept of creating the smallest, most green preference tree. ID3 makes use of the statistics gain of every characteristic for the development choice tree. The talents having a very nice benefit can pick distributed statistics data. ID3 the set of rules has a few drawbacks, which includes for some time, statistics may be over-categorized, and simplest one component at a time is taken into consideration for creating decision tree. Most effective characteristic at a time is examined so that it will make a choice, and it does longer deal with the non-stop characteristic along with the hidden charge for making a tree. This set of policies turned into design-based totally on the requirements of Occam's razor, thoughts of making the minimum, maximum green desire tree. ID3 makes use of the information gain of every function for the development choice tree. The talents having a very fine advantage can pick out for dividing information and records. ID3 the set of guidelines has disadvantages, which include for a while, statistics may be over-labelled, best one characteristic at a time is taken into consideration for a decision tree. The handiest one characteristic at a time is examined with the purpose to make a selection, and it does now not cope with the non-stop characteristic as well as the missing price for creating a tree (Neeraj Bhargava *et al.*, 2017).

12.4.3   C 4.5

C4.5 is the expansion of the ID3 sets of rules and an arithmetical classifier. It overcomes the difficulties related to the ID3 calculation like adapting to non-infer records and missing qualities. It follows the identical gadget as ID3 for specific records and uses the factor ratio technique for numerical kind of records.

12.5   Data Mining Model for Detecting the Attacks

In this method of detecting attacks KDD99 data set is used. It contains 42 features includes class column which has general and attacks type data. This model is goes through two different phases.

Inside the first section, network facts are pre-processed to convert the discreet form of information into integer kind records and after conversion; a choice tree is constructed with the assist of the facts set which is pre-processed. This pre-processed information set is sufficient in a position to differentiate the information of well-known records and assault records inside the toddler node of the tree.

Another phase is known as "Detection Phase". This section analyses the assault related to its class and its own wide variety of prevalence. It identifies noisy records or lacking facts which is called attacks (described in Figure 12.11).

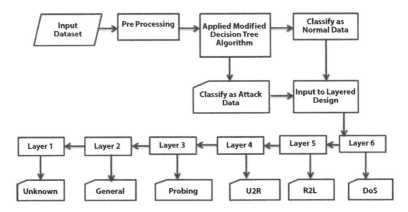

Figure 12.11 Model of modified decision tree algorithm.

12.5.1   Framework of the Technique

The model which is introduced is an enhanced model of the sets of rules of C4.5 tree which cope with both non-linear and specific information concurrently for distributed dataset as every day and fraud at leaf degree. In simple C4.5, the set of data that is dataset needs a shorter layout, and it manages unique and non-stop records one after the opposite that is a time-consuming method, and additionally deciding on the breakup value is an essential difficulty in making a decision tree.

Via specializing in such situations, we've changed some instances in our introduced model, like in preference to managing both express and non-prevent information one after the other, we convert the explicit records to continuous facts in pre-processing, and with none shoring the dataset we immediately practice the set of rules for classification. For dividing capabilities. The steps of the algorithm are as follows:

Algorithm
Input: Any Random Dataset (D)
Total numbers of samples in D: N_R
Number of unique element: N_U
Columns in dataset: C_D
Distinct value present in C_D: V
Element in C_D: D_i
Total number of unique element in N_U: T_i
Output: Classified Data
Begin
Step 1. If input dataset has similar type of class, then
        Leaf ← Class_Name

Step 2. If the particular class is present in input data
        Leaf ← Histogram

Step 3. Entropy of Dataset En $(N_R) = \sum_{i=1}^{UC} freq \frac{(Ti, NR)}{|NR|} * \frac{\log freq(Ti, NR)}{|NR|}$

Step 4. Information of every attributes $(\text{Info}_{Att}(N_R) = \sum_{i=1}^{V} |\frac{Di}{NR}| * En(Di)$

Step 5. Information Gain (IG (N_R)) = En (N_R)-Info$_{att}(N_R)$

Step 6. Division Information $(\text{Div}\_\text{Info}(N_R)) = \sum_{i=1}^{V} \frac{|Di|}{NR} * \log_w \frac{|Di|}{NR}$

Step 7. Gain Ratio $(G_{Ratio}(N_R)) = \frac{IG}{DIv\ Info(NR)}$

Step 8. Verdict Node ← Absolute gain ratio attributes
Step 9. Divide values ← Means
        Left Subset ← (Dataset< Divide values)
        Right Subset ← (Dataset>Divide Values)
Step 10. Repeat step 1 to 9 on each subset produced with the aid of dividing the set into attributes and inserts the nodes of the one as a descendant of the parent node.
End

After the successful production of the selection tree, we need to classify the fraud regular with its class, consisting of dos or U2R, R2l, or probe kind (shown below in Figure 12.12).

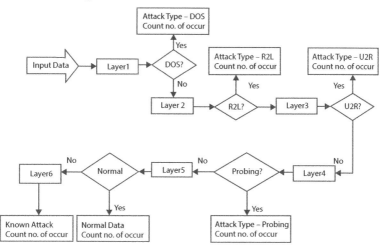

Figure 12.12 Stream of attack detection.

The enter dataset tests experience six levels in which at each confirmation the examples are when contrasted in resistance with the fraud style type. Each fraud magnificence contains attack name related to that magnificence. Tests results of each layer displays the variety of patterns related to its magnificence. Let us assume that pattern is converted into the layer structure, then increments the number of the prevalence of R2L type if no then test whether or not it's far U2R the type after which probing kind (Singh Rathore *et al.*, 2020). If the sample no longer belongs to these four attack kinds then test whether or not or not it is an everyday type record or now not. If sure then increase the range of occurrence of the ordinary kind, and if it does no longer belong to a person of those classes, then make it as unknown assault kind and be aware the range of occurrences. Ultimately, the end end result suggests the extensive form of samples found in each elegant kind. In each layer, the facts are filtered to their suitable class type. The category of the glide of attack is shown in parent 11 assault kind with its number of occurrences.

12.6   Conclusion

A Decision tree is one of the great and well-known strategies for detecting structures. This makes the correct choice of what is in community or not in the community internet site online site visitors' facts is both an attack and ordinary records. The introduced model creates a decision tree with the help of the benefit ratio and the geometric mean for splitting the dataset. It additionally efficiently identifies the specific sorts of fraud items within the dataset with the identification of unknown statistics. Result of the proposed version is in comparison with distinctive DT strategies like cart and ID3 extension (C4.5) with the help of the KDD cup-99 dataset and the proposed version gives 99% accuracy for assault identification with lesser time. The advantages of the proposed model over C4.5 are the risk to gain an immoderate detection fee over diverse sorts of fraud with tons of a good deal fewer mistakes rate and time. The destiny paintings are to check the overall overall performance of this version over a big dataset and additionally to cope with the type of unknown fraud in an automated control system.

References

A. S. Georghiades, P. N. Belhumeur and D. J. Kriegman, "From few to many: illumination cone models for face recognition under variable lighting and pose," in *IEEE Transactions on Pattern Analysis and Machine Intelligence*, vol. 23, no. 6, pp. 643–660, June 2001.

Bharat Singh, Ravinder Singh and Pramod Singh Rathore. Randomized Virtual Scanning Technique for Road Network. *International Journal of Computer Applications* 77(16):1–4, September 2013.

C. Liu and H. Wechsler, "Evolutionary pursuit and its application to face recognition," in *IEEE Transactions on Pattern Analysis and Machine Intelligence*, vol. 22, no. 6, pp. 570–582, June 2000.

Frank Y. Shih, Chao-Fa Chuang, Patrick S. P. Wang, "Performance Comparisons of Facial Expression Recognition in Jaffe Database," *International Journal of Pattern Recognition and Artificial Intelligence,* Vol. 22, No. 3 (2008) 445–459.

G.R.S Murthy R.S.Jadon "Effectiveness of Eigenspaces for Facial Expressions Recognition," *International Journal of Computer Theory and Engineering,* Vol. 1, No. 5, December, 2009, pp. 1793–8501.

J. C. Caicedo and S. Lazebnik. Active object localization with deep reinforcement learning. In *ICCV*, pp. 2488–2496, 2015.

Jyh-Yeong Chang and Jia-Lin Chen, "Automated Facial Expression Recognition System Using Neural Networks" *Journal of the Chinese Institute of Engineers,* Vol. 24, No. 3, pp. 345–356 (2001).

N. Bhargava, S. Dayma, A. Kumar and P. Singh, "An approach for classification using simple CART algorithm in WEKA," *2017 11th International Conference on Intelligent Systems and Control (ISCO), Coimbatore,* 2017, pp. 212–216. doi: 10.1109/ISCO. 2017. 7855983

Naveen Kumar, Prakarti Triwedi, Pramod Singh Rathore, "An Adaptive Approach for image adaptive watermarking using Elliptical curve cryptography (ECC)", *First International Conference on Information Technology and Knowledge Management,* pp. 89–92, ISSN 2300-5963 ACSIS, Vol. 14, 2018, DOI: 10.15439/2018KM19

P. S. Rathore, A. Chaudhary and B. Singh, "Route planning via facilities in time dependent network," *2013 IEEE Conference on Information & Communication Technologies,* Thuckalay, Tamil Nadu, India, 2013, pp. 652–655. doi: 10.1109/ CICT.2013.6558175

Pramod Singh Rathore, "An adaptive method for Edge Preserving Denoising, International Conference on Communication and Electronics Systems, Institute of Electrical and Electronics Engineers, PPG Institute of Technology (2017). *Proceedings of the 2nd International Conference on Communication and Electronics Systems (ICCES 2017)*: 19–20 October 2017.

Neeraj Bhargava, Abhishek Kumar, Pramod Singh, Manju Payal, December 17 Volume 3 Issue 12, "An Adaptive Analysis of Different Methodology for Face Recognition Algorithm", *International Journal on Future Revolution in Computer Science & Communication Engineering (IJFRSCE),* pp. 209–212, 2017.

Singh Rathore, P., Kumar, A., & Gracia-Diaz, V. (2020). A Holistic Methodology for Improved RFID Network Lifetime by Advanced Cluster Head Selection using Dragonfly Algorithm. *International Journal of Interactive Multimedia and Artificial Intelligence,* 6 (Regular Issue), 8. http://doi.org/10.9781/ijimai. 2020.05.003

13

A Maize Crop Yield Optimization and Healthcare Monitoring Framework Using Firefly Algorithm through IoT

S K Rajesh Kanna[1]*, V. Nagaraju[1], D. Jayashree[1], Abdul Munaf[2] and M. Ashok[3]

[1]*Professor, Rajalakshmi Institute of Technology, Chennai, India*
[2]*Associate Professor, Rajalakshmi Institute of Technology, Chennai, India*
[3]*Assistant Professor, Rajalakshmi Institute of Technology, Chennai, India*

Abstract

The major challenge faced by agriculture-related organizations is to identify the optimal crop that will yield higher profit based on the dynamically changing climatic conditions. The prediction of the optimal crop includes production of the crop, marketing, supply chain, storage, transportation, etc., along with the constraints and satisfying the risk associated with it. In this current research, firefly algorithm has been used for optimizing maize crop yield by considering the various constraints and risks. This research investigates the development of new firefly algorithm module for predicting the optimal climatic conditions and predicts the crop cultivation output. As the pre-processing, the maize crop cultivation data for 96 months have been collected and provided as response to Minitab software to formulate the relational equation. The collected data have been stored in the cloud using IoT and the cloud has to be updated periodically for obtaining accurate results from the algorithm. This equation has been used fitness function to deliver precise forecast of crop yield. The considered noteworthy variables are average amount of rainfall, irrigation facility and atmospheric air temperature to identify best combination which can yield higher maize crop cultivation. The performance of the developed module has been found to be satisfactory and can be used for the prediction of the other crops also.

Keywords: Crop yield prediction, internet of things, fire fly algorithm, maize crop

Corresponding author: skrkanna@gmail.com

Neeraj Bhargava, Ritu Bhargava, Pramod Singh Rathore, and Rashmi Agrawal (eds.) Artificial Intelligence and Data Mining Approaches in Security Frameworks, (229–246) © 2021 Scrivener Publishing LLC

13.1   Introduction

Crop yield profit forecasting has been a puzzling issue for agricultural-related organizations and farmers to solve, as the crop yield depends on the multifaceted collaboration among soil, climate, air, rainfall, irrigation, water, humidity and the type of crops cultivate in it, i.e., the proper combination of the bio-eco system parameters required for best yield of the crop. On the other hand, these parameters are dynamically varying and difficult to predict in advance and any deviation in the prediction results in lower profit. So a need arises for a comprehensive model to predict and identify the optimal crop that will yield the better cultivation and in turn the profit, which can be possible through the mathematical and engineering models (De la Rosa *et al.*, 1989). So many researches have been carried out by the food- and agriculture-related researchers in crop forecasting and in the art of predicting crop yields. Also it is commonly accepted by the researchers that the prediction and optimal identification of the crop production before the harvest should take place in advance. In contrast, optimal identification of the crop and forecasting philosophy have been a difficult task as it involves several categories of data gathering from diverse sources like meteorological data, agronomic data, soil strength–related data, water availability data, type of crop data, agricultural statistics, climatic data, humidity data, sunlight intensity data, rainfall and water supply data. Even though several indices for these variables have been derived in determining the crop yield, the performance of the responses have not been at a satisfactory level (Bharat Singh *et al.*, 2015). So the requirements of a model to identify the optimal crop and the conditions that will yield the maximum profit have emerged.

Timely and accurate identification of optimal crop yield are essential for maximum crop cultivation, storage, transportation and marketing that will generate maximum profit. So it became essential to make the model understand the stochastic behavior of crop yield from the previous data at all levels (Doraiswamy, Paul C., 2007). So in this research, data for the past 96 months have been collected from the famers of the Karur village in the Namakkal district, Tamilnadu, India. Also required is timely evaluation and identification of potential crops that will yield maximum profit, and in turn may have an economic impact on the agricultural products in the market (Easterling, W.E., *et al.*, 1996). So the developed model can give prediction of the optimal crop that will yield maximum profit month-wise and year-wise.

The researchers are also keen on developing models that can work all over the world based on the geo-location data and agronomic data, which can forecast the finest yield using approaches like multiple linear regression

(Gerpacio, Roberta V., 2007) Artificial neural network, Tabu search (Gommes, Rene, 2001), etc. So the main aim of this research is to use the cloud databank by means of the Internet of Things (IoT) for storing previous records and to store the current data for the identification and prediction of the optimal crop that will yield maximum profit. The main end users of the developed model are the local farmers who are directly involved in crop cultivation. The decision will be taken in line with the experience, developed model outcome and the recommendations provided by the Ministry of Agriculture. At the farm level, the developed model should have the ability to identify the controllable shortfalls and by satisfaction, will maximum profit (Bharat Singh *et al.*, 2013). So the developed model should be robust and able to accommodate all the constraints and objectives simultaneously to provide a feasible solution. Hence in this investigation, firefly algorithm has been cast off to recognize the ideal crop that can yield extreme revenue.

13.2   Literature Survey

Previously, simulation and regression were the two main models for predicting the optimal crop (Horie, T., Yajima, 1992). Simulation models are characterized through scientific relationships and the accuracy of the result depends on the available data, but the agricultural data may be sparse and incomplete, (A. Kumar, 2020; Neeraj Bhargava, Pramod Singh *et al.*, 2017) so the simulation model might not be suitable for the application. On the other hand, regression models have been used in large scale (House, C.C., 1979). Some of the researchers used the multiple regression models to predict the agro meteorological crop yield predications (Kantanantha Nantachai, 2009). Also, many new heuristic models are used to predict the crop yield by the researchers (Lansigan *et al.*, 2010). Some of the researchers specifically identify the parameters affecting the crop yield and optimize those parameters such as soil properties (N. Bhargava *et al.*, 2017; Lawless *et al.*, 2006).

Even though many researches have been carried out, most of them have not accommodated multi-constraint and multi-objective problem with alternate solutions by controlling the variables. So the only solution can be the use of evolutionary algorithms. Among the various mimic algorithms, firefly algorithm have been identified as the best algorithm that will yield better results. Apart from the weather conditions, the crop cultivation also depends on the parameters such as diseases, pest attacks, planning of harvest operation, etc. So the optimization models needs effective management of these factors also (Malaylay, R.A., Munoz *et al.*, 2011).

Some of the researchers considered the seed, fertilizer, agrochemical and agricultural machineries for crop production estimates (Marinkovic *et al.*, 2010). Some of the researchers considered the erratic and unpredictable weather conditions such as storms which affect the crop growth yields (Oludhe, Christopher, 2002). The commonly implemented models are linear regression models (Raorane, A.A., Kulkarni, 2017), nonlinear regression models (S K Rajesh Kanna, 2017) and polynomial regression models (S.K. Rajesh Kanna, 2014). Researchers also considered climatic changes to simulate corn performance yield (S.K. Rajesh Kanna *et al.*, 2015). Multiple linear regression equation had also been constructed to correlate the yield and response variables (S.K. Rajesh Kanna, 2017). Many researchers deliberated over rainfall data, surface temperature, moisture and vegetation index of various states and collected data for years (S.K. Rajesh Kanna *et al.*, 2015).

(Lawless and Semenov, 2006) conducted experimentation by analyzing six different sites having diverse climatic changes in Europe and New Zealand (Savin, I, 2010) focused on nitrogen application and weed control strategies using ANN and Decision Trees. But the decision tree model has limitations like unstable results, less number of variable considerations, etc. So in this research firefly algorithm have been used to optimize and identify the best crop that will yield better profit and maximum cultivation (Singh Rathore *et al.*, 2020).

13.3   Experimental Framework

After the review on the existing methodologies related to crop yield, this section deals with the implementation of firefly algorithm for the identification of optimal crop for yielding maximum profit. The entire research work has been divided into different sub-categories such as Data collection, Data preprocessing, IoT and cloud database formulation, Initial considerations and assumptions, Formulation of relational equation, Implementation of firefly algorithm, Validation and Updating cloud. The experimental framework architecture is shown in Figure 13.1. The first stage is the data collection stage and this experimental study had been carried out in the villages near Karur, Namakkal District, Tamilnadu, India. The maize crop yielding–related data for consecutive cropping seasons of 96 months have been collected from the farmers and the agriculturalist.

The data have been collected from one to one surveying, by circulating the forms, oral communications through question and answer mode, brainstorming with the framers, one to one meeting, data sheet collection, Google forms, etc. The data collected are the yield per hectare. After collecting the data, the collected data have to be validated for correctness, so that the

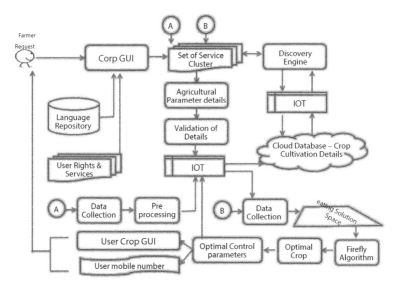

Figure 13.1 The Architecture of the experimental framework.

developed models can produce the desired performances. The scatter plot for the collected data have been formulated and plotted. The plotted scatter plot is given in Figure 13.2. The second stage is the data preprocessing stage. In general, the real-world data collected by surveying are often incomplete, redundant, inconsistent, and lacking in certain aspects. So a need arises for the data to be filtered for completeness. It is a rule of thumb that the complete data will give accurate results and the improper data results in the near optimal solutions and consumes an additional time period.

The steps followed in this research for filtering the corrected data are as follows:

(a) Data series with missing values are omitted from the dataset,

(b) Data with sudden extreme peak values are rounded to the nearby average values,

(c) Filling in missing values based on the surrounded nine data values,

(d) Smoothing the noisy data and

(e) Identification and elimination of the inconsistent data.

The third stage is the IoT and Cloud database storage. Because of various advantages such as the developed model has to identify the optimal values of various segments, data should be available in the common space, data of the crop yield have be updated periodically for accurate prediction

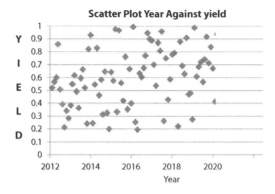

Figure 13.2 Scatter plot of the collected data.

in the future, data can be interpreted by the agriculturalist for other predictions and statistics. So the only possibility is to append of the data in the cloud storage and reclaim for the further processing.

The fourth stage is the initial considerations for the application. In order to shrink the computational complication, certain parameters with incomplete data and under controllable categories are assumed as follows, good quality seeds, no storms, sufficient fertilizers, proper communication and technology, uniform temperature, sandy clay loam soil, weed and pest control, plowing and harrowing.

The fifth stage is the formulation of the empirical equation representing the relationship between the input and the output parameters. So optimizing different units simultaneously took more complications and thereby parameters are optimized and given as the input to Minitab software, the obtained relational equation is given in Equation 13.1.

$$Y(x) = A - B$$
$$A = \{(10.135 * r) + (37.291 * i) - (4.435 * t) + (12.642 * i)\}$$
$$B = \{ (14.121 * r * r) + (1.823 * i * i) + (7.078 * t * t) + (0.238 * i * r)$$
$$+ (23.138 * r * i) + (3.938 * r * t) - (4.337 * r * i) - (22.867 * i * t)$$
$$+ (8.981 * i * r) + (7.324 * t * i) + 189.578 \qquad (13.1)$$

Whereas, 'r', 'i' and 't' represent the rainfall, irrigation and temperature parameter, respectively, for the corresponding period of time.

Obtained equation has been used as objective function or fitness equation for firefly algorithm with the objective to maximize the value. Y(x) denotes the amount of yield for maize crop for the time period 'x' for that location. (N. Bhargava *et al.*, 2017) Obtained yield value has been in the range of 0 to 1 and that has been converted to user understandable percentage of yield format.

The sixth stage is implementation of firefly algorithm for identifying optimal crop and optimal climatic parameters for yielding maximum profit. The initial stage is to establish the number of fireflies and construction of search space that will be used in the search (Shibayama, M., 1991). The performance of the firefly algorithm is influenced by control factors. So the number of fireflies can be calculated using the Equation 13.2.

$$nof = Cint \left(\sqrt[n]{no. \ of \ months \times no. \ of \ control \ parameters} \right) \quad (13.2)$$

All the virtual fireflies search the optimal solution based on the collective knowledge obtained by the elder fireflies to the best location and paths. The fireflies select their path based on the attractiveness or light intensity of the other fireflies (Uno, Y., et al., 2005). The attractiveness or light intensity "I" of a firefly can be calculated using the Equation 13.3.

$$I(r) = (I_j/r_{ij}^2) \quad (13.3)$$

Whereas, 'r' is the distance between the ith and j^{th} firefly and $I_j = F(X)$, the objective function. Each firefly has a certain attractiveness coefficient value 'β' and can be calculated from the Equation 4.

$$\beta = \beta_0 \, e^{-\gamma r^2} \quad (13.4)$$

Whereas, 'β_0' is the primary firefly attractiveness value and 'γ' is the light absorption coefficient [24]. The movement of a firefly can be calculated using the Equation 13.5.

$$X_j = X_i + \beta_0 \, e^{-\gamma r^2} (X_j - X_i) + \alpha \varepsilon^i \quad (13.5)$$

Whereas, it comprises its previous position, attraction of the firefly and randomization parameter to prevent stagnation (Walker, G.K., 1989). If the calculated newer position having higher attractiveness then the existing, then that firefly will move to the identified new position and the movement has to continue till the predefined nodes are reached, i.e., three nodes in this case.

In every iteration, the best path has to be stored and the same procedure has to be repeated till the number of iterations reached 100 shown in Figure 13.3 (Wilcox, A., et al., 2000). The seventh stage is the validation of the developed algorithm and developed algorithm has been tested for the month of February month and the month of March. The algorithm for the developed firefly is shown in Figure 13.4.

Figure 13.3 Output GUI with sample output.

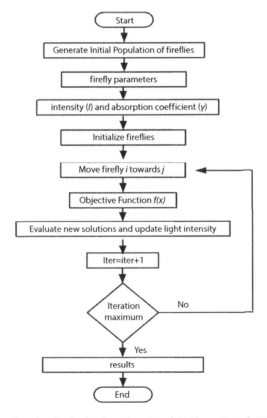

Figure 13.4 Flowchart for the firefly algorithm (*Singh Rathore, P. et al. 2020*).

13.4   Healthcare Monitoring

The developed cloud has also been used to monitor the healthcare of the farmers and the same developed module also suggests the recommendations based on the data obtained. In the healthcare monitoring the blood pressure of the farmers has been monitored, and the Android mobile app GUI is shown in Figure 13.5.

The heart rates of the farmers have been monitored by the developed IOT-based system and the developed GUI is shown in Figure 13.6.

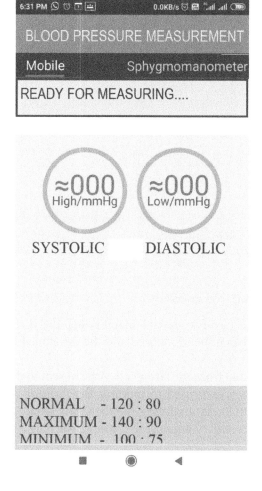

Figure 13.5 Blood pressure measuring GUI.

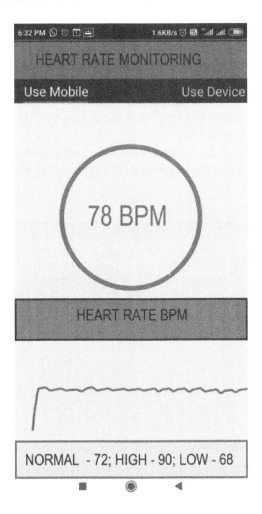

Figure 13.6 Heart Rate measurement GUI.

In general, the farmers are not maintaining the diet and timing, so the oxygen level has to be monitored and the developed GUI is shown in Figure 13.7.

The data collected from the farmers have been sent to the cloud through the same wireless sensor used to monitor the farming conditions (Zhu, Wenjun, *et al.*, 2013). By analyzing the data, The cloud gives warning to the farmers for necessary actions; the sample warning message is shown in Figure 13.8.

The developed module also stores the collected information in the cloud and can be interpreted at any time. The sample stored data is shown in Figure 13.9.

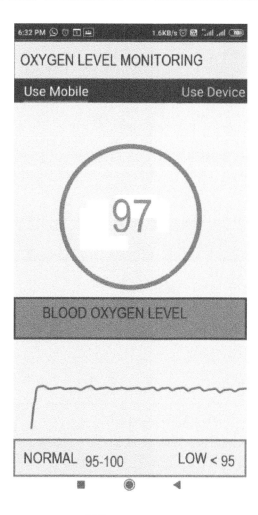

Figure 13.7 Oxygen measurement GUI.

Thus the developed IOT module has been used for collecting the agricultural data and also to monitor the healthcare of the farmers. Thus the objective is not only to enhance the cultivation profit but also the health of the farmers. For the experimental work, thinks lab cloud has been used and for commercial use, a separate cloud database has to be used. Also the data can be sent to the nearby primary health centers and to the relatives for necessary meditations; thereby the health of the farmers is enhanced. This type of application is essential for the rural areas of developing countries, where the farmers do not have much knowledge about their health.

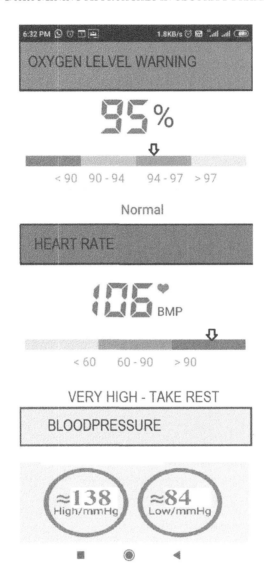

Figure 13.8 Warning message GUI.

13.5   Results and Discussion

The maize farm area selected for experimental study was Karur village, Tamilnadu state, south India. Data gathered from farmers and agriculturalists were segregated into primary and secondary data. Major influencing data such as maize output, rainfall, input cost, irrigation facility and

Figure 13.9 Cloud database value.

temperature as primary data and data like yielding season, demand, fertilizers, seed quality, etc., as secondary data.

As data collected over a very long range of years, the average value of the data from 30 maize farmers' data have been taken into consideration. Contradictory data and incomplete data were removed during preprocessing.

In the developed firefly algorithm model, inputs $I = \{X_1, X_2, X_3\}$, whereas $X_1 = \{$Rainfall $R_1, R_2, R_3, \ldots R_n\}$ in centimeters, $X_2 = \{$temperature $T_1, T_2, T_3, \ldots T_n\}$ in centigrade and $X_3 = \{$Irrigation facility Available, not available$\}$ in Boolean. The yield function shown in the Equation 1 is maximization function and in-turn maximum profit. By simulating firefly model, the results shows that highest yield of maize will be 68 bags/hector and the lowest yield will be 18 bags/hector for the year 2020. The sample output of the developed model is shown in Figure 13.3.

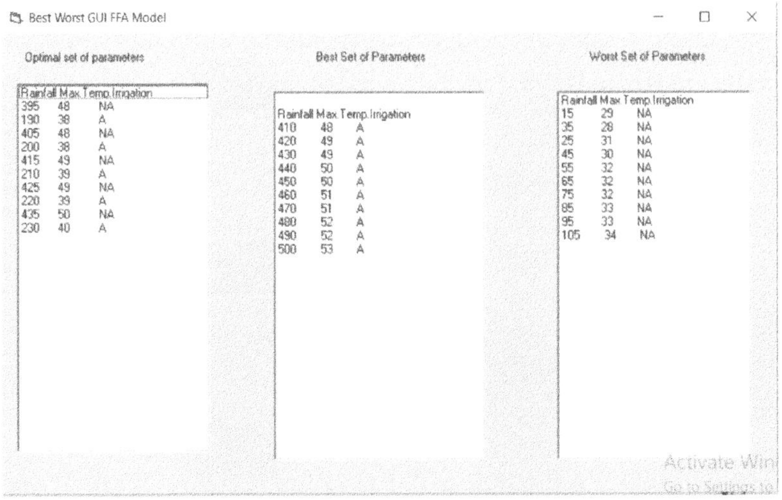

Figure 13.10 GUI for optimal, best and worst parameters.

Each farmer has been given a unique ID and farmers have to use it for data entry and for predications. In geo-location mode, farmers are allowed to select region and location. In this module, only one location, Karur village has been considered. Consequently, this established model helps the farmers to foresee and recognize an appropriate period for maximum maize cultivation and this in turn results in high profitability. Also the developed model forecast the optimal parameters, which are better for cultivation to the farmers. Thus farmers can proceed with suitable judgment to maximize their revenue. Sample output section is specified in Figure 13.10.

A total of 1,103 facts were surveyed for 16 parameters from 2012 to 2019. Upon preprocessing, three datasets have been found complete and having major influence over maize crop cultivation with 288 data points. Missing data points and parameters were considered for future scope.

13.6   Conclusion

In this paper, new research possibilities on using firefly algorithm for identifying and predicting optimal parameters for maize crop yielding application in maximizing profit have been successfully implemented using IOT technology. Three climatic variables such as temperature, rainfall, and irrigation facility have been collected for 96 months and preprocessed for correctness. It has been used to generate relational formula between variables using

Minitab software and has been used as fitness function for firefly algorithm. The outcome of firefly algorithm have been validated with real-time data and it has been found that the results are in track with existent time data. Thus the established unit can be used for prediction of process parameters for other crops and for various geographical locations. The main advantage of the developed model is that the data storage is in the cloud and the cloud will be updated continuously for accurate data. Also cloud data can be used for the other statics by the government and researchers. The developed IOT module also multitasked for the monitoring of the health of the farmers.

References

De la Rosa, D., Cardona, F., Almorza, J. (1981), "Crop Yield Prediction Based on Properties of Soils in Sellilla, Spain," *Journal of Geoderma*, 25 (3-4), 267–274.

Gommes, Rene, "An Introduction to the Art of Agrometeorological Crop Yield Forecasting Using Multiple Regression," 2001.

Horie, T., Yajima, M., Nakagawa, H. (1992), "Yield Forecasting," *Agricultural Systems*, 40 (1-3), 211–236.

Bharat Singh, Ravinder Singh and Pramod Singh Rathore. Randomized Virtual Scanning Technique for Road Network. *International Journal of Computer Applications* 77(16):1–4, September 2013.

Singh Rathore, P., Kumar, A., & Gracia-Diaz, V. (2020). A Holistic Methodology for Improved RFID Network Lifetime by Advanced Cluster Head Selection using Dragonfly Algorithm. *International Journal of Interactive Multimedia and Artificial Intelligence*, 6 (Regular Issue), 8. http://doi.org/10.9781/ijimai.2020.05.003

N. Bhargava, S. Dayma, A. Kumar and P. Singh, "An approach for classification using simple CART algorithm in WEKA," *2017 11th International Conference on Intelligent Systems and Control (ISCO), Coimbatore*, 2017, pp. 212–216. doi: 10.1109/ISCO. 2017. 7855983

Naveen Kumar, Prakarti Triwedi, Pramod Singh Rathore, "An Adaptive Approach for image adaptive watermarking using Elliptical curve cryptography (ECC)", *First International Conference on Information Technology and Knowledge Management* pp. 89–92, ISSN 2300-5963 ACSIS, Vol. 14, DOI: 10.15439/2018KM19

Prof. Neeraj Bhargava, Abhishek Kumar, Pramod Singh, Manju Payal, "An Adaptive Analysis of Different Methodology for Face Recognition Algorithm", *International Journal on Future Revolution in Computer Science & Communication Engineering (IJFRSCE)*, Volume 3, Issue 12, December 17 2017, pp. 209–212.

Doraiswamy, Paul C., "Operational Prediction of Crop Yields Using MODIS Data and Products", *International archives of photogrammetry remote sensing and spatial information sciences*, 2007.

Easterling, W.E., *et al.*, "Improving the Validation of Model-Simulated Crop Yield Response to Climate Change: An Application to the EPIC Model.", *Inter-Research Science Center*, Vol. 6, No. 3 (June 13, 1996), pp. 263–273.

Gerpacio, Roberta V., Pingali, Prabhu L., "Tropical and Subtropical Maize in Asia Production Systems, Consraints, and Research Priorities", Maize Production Systems Papers 56107, CIMMYT: International Maize and Wheat Improvement Center. 2007, DOI: 10.22004/ag.econ.56107

A. Kumar, P.S. Rathore, A. Dubey, R.A. Grawal. (2020) "Low-Power Traffic Aware Emergency based Narrowband Protocol with holistic Ultra Wideband WBAN approach in biomedical application" Accepted in *Ad Hoc & Sensor Wireless Networks Journal*, ISSN: 1551-9899 (print) ISSN: 1552-0633 (online).

House, C.C. (1979), "Forecasting Corn Yields: A Comparison Study Using 1977 Missouri Data," Unnmbered report, Statistical Research Division, United States Department of Agriculture.

Kantanantha Nantachai, "Crop Decision Planning Under Yield and Price Uncertainties", H. Milton Stewart School of Industrial and Systems Engineering, Georgia Institute of Technology, 2007.

Lansigan, *et al.*, "Analysis of Climatic Risk and Coping Strategies in Two Major Corn Growing Areas in the Philippines," 2010.

Pramod Singh Rathore, "An adaptive method for Edge Preserving Denoising, International Conference on Communication and Electronics Systems, Institute of Electrical and Electronics Engineers & PPG Institute of Technology (2017). *Proceedings of the 2nd International Conference on Communication and Electronics Systems (ICCES 2017)*: 19-20, October 2017.

Lawless, Conor and Mikhail A. Semenov, "Assessing Lead-time for Predicting Wheat Growth Using A Crop Simulation Model", *Agricultural and Forest Meteorology* 135(1–4):302-313, DOI: 10.1016/j.agrformet.2006.01.002, 2006

Lingaraj N, S K Rajesh Kanna, G. Suresh, P. Sivashankar, "Implementation of Firefly Algorithm for the 1D Bin Packing Problem with Multiple Constraints", *International Journal of Research*, 2019, Jan, Vol 8, Issue 1, pp. 1188–1194.

Malaylay, R.A., Munoz, M.J., Paglinawan, K.L., "Proposed Crop Decision Planning Model for Yellow Corn Production in Quezon Province for the Years 2011–2015," February, 2011.

Neeraj Bhargava, Pramod Singh, Abhishek Kumar, Taruna Sharma, Priya Meena, December 17 Volume 3 Issue 12, "An Adaptive Approach for Eigenfaces-based Facial Recognition", *International Journal on Future Revolution in Computer Science & Communication Engineering (IJFRSCE)*, pp. 213–216.

Marinkovic, *et al.*, "Data Mining Approach for Predictive Modeling of Agriculture Yield Data," 2010.

N. Bhargava, A. Kumar Sharma, A. Kumar and P. S. Rathoe, "An adaptive method for edge preserving denoising," *2017 2nd International Conference on Communication and Electronics Systems (ICCES), Coimbatore*, 2017, pp. 600–604, doi: 10.1109/CESYS.2017.8321149.

N. Bhargava, S. Sharma, R. Purohit and P. S. Rathore, "Prediction of recurrence cancer using J48 algorithm," *2017 2nd International Conference on Communication and Electronics Systems (ICCES), Coimbatore*, 2017, pp. 386–390, doi: 10.1109/CESYS.2017.8321306.

Oludhe, Christopher, "Deterministic and Probabilistic Prediction Approaches in Season to Inter-annual Climate Forecasting," 2002.

Raorane, A.A., Kulkarni, R.V., "Data Mining: An Effective Tool for Yield Estimation in the Agricultural Sector." 2012, *Agricultural Sciences*, Vol. 8, No. 5, May 22, 2017.

S K Rajesh Kanna, "Application of Firefly Algorithm and Its Parameter Setting for Job Shop Scheduling", *International Journal of Advanced Research in Management, Architecture, Technology and Engineering*, March 2017, Vol. 3, Issue 9, pp. 3–6.

S.K. Rajesh Kanna, A.D. Jaisree, "Forecasting the Optimal Yielding Period and Profitability of Maize Cropping System Using Genetic Algorithm", *European Journal of Applied Engineering and Scientific Research*, Scholars Research Library., June, 2014, vol. 3, issue 2, pp. 13–18.

S.K. Rajesh Kanna, Jaisree, "Application of Firefly Algorithm for Optimizing Bevel Gear Design Problems in Non-Lubricated Condition", *International Journal Of Research In Computer Application & Management*, Nov 2015, Vol. 5, Issue 11, pp. 26–29.

S.K. Rajesh Kanna, K.C. Udaiyakumar, "A Complete Framework for Multi-Constrained 3d Bin Packing Optimization Using Firefly Algorithm", *International Journal of Pure and Applied Mathematics*, 2017, Vol. 114, Issue 6, pp. 267–282.

S.K. Rajesh Kanna, S. Stella, "Optimizing Maize Crop Yielding and Cropping System Using Tabu Search Approach", *International Journal of Scientific Engineering and Applied Science*, Sep. 2015, Vol.1, Issue 6, 146–151.

Savin, I., "Crop Yield Prediction with SPOT VGT in Mediterranean and Central Asian Countries," 2010.

Shibayama, M. (1991), "Estimating Grain Yield of Maturing Rice Canopies Using High Spectral Resolution Reflectance Measurements," *Remote Sensing of Environment*, 36(1), 45–53.

Uno, Y., *et al.*, "Artificial Neural Networks to Predict Corn Yield from Compact Airborne Spectographic Imager Data", *Computers and Electronics in Agriculture*, vol. 47, issue 2, https://doi.org/10.1016/j.compag.2004.11.014 2005.

Walker, G.K. (1989), "Model for Operational Forecasting of Western Canada Wheat Yield," *Agricultural and Forest Meteorology*, 44(3-4), 339–351.

Wilcox, A., *et al.*, (2000), "Factors Affecting the Yield of Winter Cereals in Crop Margins," *Journal of Agricultural Science*, 135(4), 335–346.

Zhu, Wenjun, *et al.*, "Improving Crop Yields Forecasting Using Weather Data: A Comprehensive Approach Combining Principal Component Analysis and Credibility Model.", *Second International Agricultural Risk, Finance, and Insurance Conference (IARFIC)*, June 16-18, 2013, Vancouver, BC, Canada.

14

Vision-Based Gesture Recognition: A Critical Review

Neela Harish*, Praveen, Prasanth, Aparna and Athaf

Assistant Professor, Student, Rajalakshmi Engineering College, Chennai, India

Abstract

Interactions between a human being and a machine have become part of common and essential services. This is known as Man-Machine Interface and is divided into detection, tracking and recognition. In this study Vision-based gesture recognition has been described in detail on the basis of: Acquisition, features, classifier rate. Gestures are of two types: static and dynamic sequences, this is where vision-based techniques play a vital role. The survey on the research study on the vision-based gesture recognition approaches have been briefed in this paper. Challenges in all perspective of recognition of gestures using images are detailed. A systematic review has been conducted over 100 papers and narrowed down to 60 papers and summarized. The foremost motive of this paper is to provide a strong foundation on vision-based recognition and apply this for solutions in the medical and engineering fields. The paper outlines gaps and current trends to motivate researchers to improve their contribution.

Keywords: Machine learning, image-based gestures, ANN

14.1 Introduction

- Gestures are the expressive form of human beings to convey information. The three roles are Semiotic, Ergotic and Epistemic. The gestures are captured through camera or sensor and then processed by a computer. Research in this field has become trending in the market for application of

---

Corresponding author: neela.m@rajalakshmi.edu.in

Neeraj Bhargava, Ritu Bhargava, Pramod Singh Rathore, and Rashmi Agrawal (eds.) *Artificial Intelligence and Data Mining Approaches in Security Frameworks*, (247–260) © 2021 Scrivener Publishing LLC

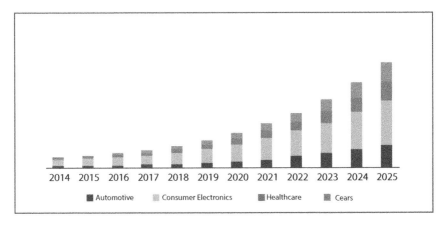

Figure 14.1 Gesture recognition product growth over Asia-Pacific regions (Source: finance.yahoo.com).

medical, virtual environment, etc. Vision-based products have a variation in cameras and their features: lens, aperture, resolution, shutter, battery. Figure 14.1. presents the economic product growth of Vision-based gesture recognition systems over Asia-Pacific regions obtained from a survey showing tremendous growth in recognition products since 2014 and continuing up to 2025, as it observes its applied advantages in sign language recognition. This paper is organized as follows: Section II briefs challenges in vision-based gesture recognition, Section III describes the image processing, Section IV is a literature review; Section V summarizes the existing models and Section VI presents the conclusion.

The gestures have been classified based on various categories as per the following:

- Pointing
- Semaphoric
- Pantomimic
- Iconic
- Manipulation

14.2   Issues in Vision-Based Gesture Recognition

The challenges faced by each device on Vision-based gesture recognition is categorized in three basic units (A. Kumar *et al.*, 2019) as in Figure 14.2, Figure 14.3 and Figure 14.4. The challenges could be based on the parameters of computerized system, gesture and environment situations. It may cause change in the response time, factor of cost, background illumination effects such as scaling and rotation, dataset sample size, image processing steps of segmentation and feature extraction of static and dynamic gestures. In spite of these factors a real-time image-based processing has been designed and developed for the gesture recognition (A.V. Dehankar and Dr. Sanjeev Jain, 2017).

14.2.1   Based on Gestures

(Abhishek Jain *et al.*, 2018): The gestures range from simple to complex; static to dynamic. Various problems during gesture capture are: Translation, scaling rotation of images, multiview required to obtain the gesture position in all directions: Face recognition (raise eyebrows, winking, nostrils, head shaking, expressions), Classifiers vary for type of gestures Static (NN) and dynamic (DTW).

14.2.2   Based on Performance

The device designed for Vision-based recognition has the need to satisfy some specific criteria: Response time of processing, time consumed between input to output, Speed of transmission in wireless modules, latency avoided, presentation, efficiency – How many frames of image per second can be captured? (Aruljothy. S. *et al.*, 2018). Accuracy –- Maintain robustness without changing image background. The feature extraction also estimates the performance while selecting low-level features such as edges and histograms, which led to high error rate.

14.2.3   Based on Background

Preprocessing stage is vital as it should not cause any disturbances in terms of Illumination, Occlusion, Consistent lightening required, fixed distance from the lens and objects out of focus. User requirements to wear cumbersome glove and remove accessories (Bin Feng *et al.*, 2017).

14.3   Step-by-Step Process in Vision-Based

Image are classified into vector and digital [46]. A digital image is 2-dimensional array of real number, divided into N-rows and M-column the intersection of rows and columns is known as pixels. Each pixel is black

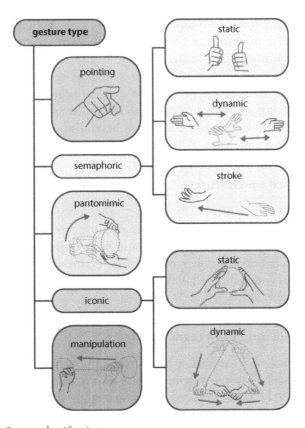

Figure 14.2 Gesture classification.

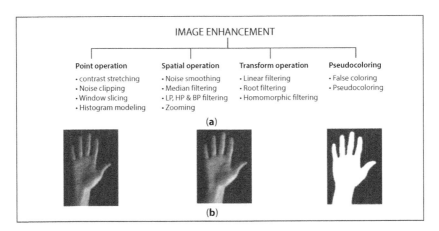

Figure 14.3 (a) Operations in Image Enhancement (b) Hand gestures in the process of hand enhancement.

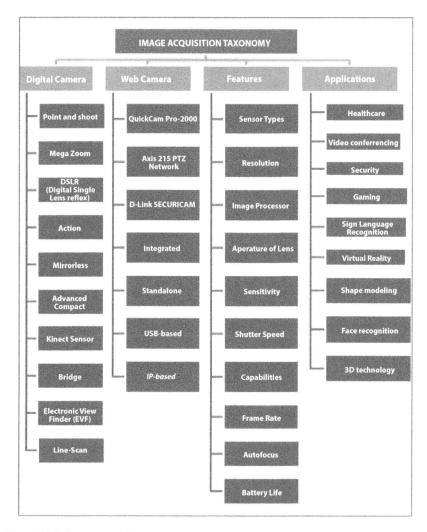

Figure 14.4 Image acquisition taxonomy.

and white normally ranging from 0 to 255, In color images pixels described by the amount of RGB with four stages in Vision-Based recognition (Dimitry Ryumin *et al.*, 2019).

14.3.1 Sensing

The combination of "illumination" source and the reflection or absorption of energy from that source by the element of the "scene" .1) single sensing element, 2) Line sensor, 3) Array sensor (Singh Rathore *et al.*, 2020).

14.3.2   Preprocessing

The aim is noise reduction, enhancement and restoration, segmentation
- Salt & pepper – randomly scattered
- Gaussian – random fluctuations added
- Speckle - random value multiplied
- Uniform – due to quantization

a) Image enhancement
   A measure of manipulating the preprocessed images to make it adjustable to other image analysis–based application. This process maintains image characteristics such as tolerated noise level and region of interest marked (Du Jiang, Zujia *et al.*, 2018).
b) Image restoration
   It enhances image with known degradation techniques. Mispresented images are reverted back to convey its original information. This transformation is done without loss to the image content as it is an objective process.
c) Image segmentation
   The division of an image into regions or categories, which correspond to different objects or parts of objects. The image pixels are modified on the basis of two parameters: discontinuity and similarity among the image pixels (Guangming Zhu *et al.*, 2018).

14.3.3   Feature Extraction

a) Feature detection refers to finding the features in an image, region, boundary.
b) Feature description assign quantitative attributes to the detector features; for example, we might detect corners in region boundary, and describe those corners by their orientation and location, both are which are quantitative attributes (Jiaqing Liu *et al.*, 2019).
c) Feature processing methods are subdivided into three principal categories depending on whether they are applicable to boundaries, region or whole images, parameter such as scale, translation, rotation, illumination and view point.
d) Some features are applicable to more than one category; feature descriptors should be as insensitive as possible to

variation in parameter such as scale, translation, rotation, illumination and view point.

e) Feature extraction is the process by which the certain feature of interest within an image are detected and represented by further processing. It is a critical step in computer vision and image processing solution it marks transition from pictorial and non-pictorial data representation.

14.4 Classification

The various classifiers (Kamal Preet Kour *et al.*, 2017) in present vision-based recognition are summarized in Table 14.1.

Table 14.1 Comparison of classifiers.

Classifier	Benefit	Limitation
K-Nearest-Neighbour - Usage of kernels and hyperplane for the categorization of 3D datasets.	Effective, Non-parametric	Memory and time establishment link takes a long time limit, which delays the classification process.
Artificial Neural Network - Nodes whose input arise outside the network are called input nodes and simply copy the value	Produce a good result in complex domains. Testing is very fast	Training is relatively slow, Empirical Risk Minimization
Support vector Machine - linear separating hyperplane	Inherent data better, independent of the dimensionality of the feature space.	Parameter tuning, Kernel Selection
Decision Tree - Flow chart type, where each node is represented as an test.	No prior knowledge is required and easy to interpret	Only one output and dependent on the data set
Bayesian Classifier - Determine the mean and covariance of the normal and abnormal functions.	The process of computing is easier	For dependent variables, more accurate results cannot be delivered

14.5 Literature Review

Digital Image Processing (DIP) is the process of digital images using various computer algorithms. This paper presents a brief overview and literature review of digital image processing techniques such as image pre-processing, image compression, edge detection and segmentation. Design and Fabrication of prior and high recognition rate products are Prototype (Liwei Yang *et al.*, 2019) which is able to automatically recognize sign language to help deaf and dumb people to communicate effectively. Convexity hull algorithm is implemented for finger point detection and number recognition. The software aims at recognizing the orientation, Centre of mass centroid and various shape-based features of hand gesture. Using 640*480 pixels with distinct features of the users may make recognition more difficult. In another approach (Maisyasenand Shaidahjusoh, 2019) This paper illustrates the recognition of hand gesture by employing an android device. The aim of the paper is to recognize 40 different images. The main characteristics are computing the centroid in the hand, presence of thumb and number of peaks in the hand gesture. Recognition is based on artificial neural networks. The android device sensing element (Webcam android application) senses the gesture and sends it as an input to the computer. The gesture is detected for edges. Edge detection is a process that aims at identifying points in a digital image at which the image brightness changes sharply or has discontinuities with three detectors Sobel, Canny, Prewitt (Mohamed A, 2019). Thinning is a morphological operation that is used to remove selected foreground pixels from binary images. It is used so that the edges are in thin lines. In this method sober edge detector is used. The detector calculates the gradient of the image intensity at every point, giving the direction of the largest possible increase from light to dark and the rate of change in that direction. The angles A and B are the two necessary angles which will be fit into the neuronal network layers. With the two angles we can exactly represent the direction of the hypotenuse from point P1 to P2 which represents the direction of a hand image an average of 77% along with higher tolerated noise levels (N. Bhargava *et al.*, 2017).

(Nagul Nagpal, 2015) proposes an effective static hand gesture recognition method by using a novel effective descriptor, named DM-based BCF which encodes hand shape information from depth maps and is compact and discriminative double-handed gestures. This system is limited to finger and hand-core occlusion make and it still needs to be evaluated in real applications. Smart (Oyebade K *et al.*, 2016) robotic shopping trolley though there are several inputs there is this feature called sign language recognition by means of video recording using Kinect 2.0 camera which process the data and gives the correct output in the form of voice recording with 95% accuracy of double-handed gestures, with only one disadvantage of higher misclassification rate. (Pichao Wang *et al.*, 2018) propose a multimodal gesture recognition method based on 3-D convolution and convolutional LSTM for gesture recognition. 3-D CNN is utilized to extract short-term spatiotemporal features from the input video followed by convolutional LSTM to learn long-term spatiotemporal features further. Spatial Pyramid Pooling is adopted to normalize the spatiotemporal features. Fine-tuning is evaluated and state-of-the-art performances on the Isolde and SKIG datasets are reported. Gesture sequences with terrible illumination cannot be well recognized in the Experiments; Fast and tiny gestures are difficult to recognize. (Rishabh Agarwal, 2016) recognized Islip former Indian sign language is recognized and for easy understanding it fed to a system which avoids the basic manual need to teach sign language, but eigen values estimation consumed time. In the latter, 24 alphabets of Indian sign language in a live video can be recognized, with a low recognition rate. In [30] hand gesture recognition system with enhanced deep learning using convolution neural network (CNN) and depth camera for human-computer interaction. It also utilizes the Kinect sensor to detect hand gestures with accuracy of 84.67% for ASL Finger Spelling with large disturbances from textures and occlusions. In this paper the three image-based depth representations as Dynamic Depth.

Image (DDI), Dynamic Depth Normal Image (DDNI) and Dynamic Depth Motion Normal Image (DDMNI) are proposed, where DDIs mainly exploit the dynamic of postures, and DDNI and DDMNI, built upon norm vectors, effectively exploit the 3D structura, limited by the objects involved in the actions with similar motion patterns are hard to distinguish in the depth maps information captured by depth maps. Two levels of image processing by means of human–computer interface; one is by Haar-like features which represents the Haar-like features are more stable in providing the correct skin color object and it easily differentiates the dark region from the skin. The other one is the ada-booster which helps in providing binary values which will be read by computer using Haar features. ASL (Saeideh Ghanbari *et al.*, 2019) proposes a waterfall architecture for combining the submodules. Human (Saransh Sharma *et al.*, 2019) interact with accessories, gestures directly handling the computer. Gyro MPU6050 acquire gestures and recognition pre-treatment to experiment uses an attitude sensor that is tied in the person's wrist. A similar review conducted on vision based was by Carlos and Robin, where some depth cameras such as Microsoft Kinect, ASUS Xtion, Mesa Swiss Ranger and

Table 14.2 Comparison of exiting vision based.

S. no.	Author	Year	Prototype proposed	Accuracy	Limitations
1.	Abhishek [2]	2018	PCA-single sign language gestures	97%	Limited to number of images and requirement of laptop
2	Alaa Barkoky [3]	2011	Thinning and wrist cropping-gestures	96.62%	High segmentation error and only numbers are recognized.
3	Amit Kumar [4]	2014	Marathi sign language-Canny algorithm	NA	Dynamic signs and facial expressions could not be recognized
4.	Boraste [10]	2015	32 sign language images are converted to text	NA	Affected by environment conditions
5.	Chao Sun [12]	2013	73 ASL signs converted with HOG and Kinect features	85.5%	High misclassification rate and time consuming.
6.	Jeevan [18]	2016	ISL gestures-SVM	93.3%	Restricted gestures to time limitation
7.	Lin [17]	2014	XBOX Kinect camera-single hand gesture	95.6%	Since images used takes around 200-600 it becomes a tedious process.
8.	Jiaqing [19]	2019	CNN with 37500pairs of Color and depth images	87.4%	cluttered backgrounds and hand self-occlusion
9	Kishore [24]	2015	ISL conversion Fourier methods	95.1%	More number of camera are used.

(Continued)

Table 14.2 Comparison of exiting vision based. (*Continued*)

S. no.	Author	Year	Prototype proposed	Accuracy	Limitations
10.	Koli [25]	2012	Single handed sign language gestures	90%	Loss of one ring results in wrong result
11.	Miao [29]	2016	ASL gestures using CNN and SVM	96.1%	Deeper CNN is expensive
12.	Oyebade [34]	2016	ASL recognition-stacked demonizing auto encoder	92.83%	Effective processing of large number of gestures

Stereo video cameras for better identification of the gestures. There are two stages: Hand localization and Gesture localization. Hand localization is based on the computer vision whereas Gesture localization is based on the machine learning. The limitation of using HMM requires the addition of bigram models and the system cannot recognize a larger vocabulary. In one approach ASL methodology, survey, result analysis, drawback was taken in (Shweta S, 2016). Surveys in (Shining Song, 2018) relegate the need, benefits of Vision-based. The comparisons of a few prototypes have been compared in Table 14.2. Also one typical personalized vision-based application has been illustrated in Figure 14.5. Subha and Balakrishnan have

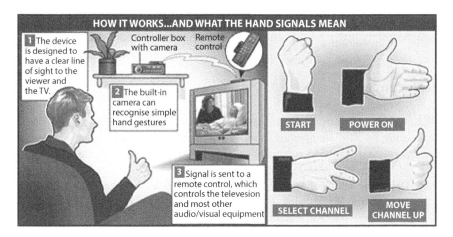

Figure 14.5 Vision controlled remote (www. zdnet.com).

designed 32 up and down positions of single-handed gesture with images. Now these 32 signs are being compared with the trained 320 images. After the extraction and threshold process are completed it is further moved to the testing phase using any kind of software, say MATLAB. Additionally, the raw data obtained through visual method (camera) need to be transmitted to multiple users, storage devices and application specific. This transmission could be made more globally with the usage of cloud computing. This enables the captured visual images to be processed into required information, of capacity up to one tetra byte. One accessing the raw data using cloud server, it provides several benefits from the user perspective: more users at a limited cost, multiple devices can be connected simultaneously, no special hardware requirement and faster access of data. But it is also limited due to reasons such as no recovery of data is available and cannot have the control over the operation of data present in cloud (Rathore, P.S. *et al.*, 2021).

14.6   Conclusion

Designed gesture recognition is to reduce the gap between disabled and normal people. Two methods are proposed: Vision and Sensor approaches. An equal amount of information and study have been going into both. In this paper, a wide variety of Vision-Based work, along with their issues, limits & successful rates are considered. Earlier models were compared with the existing vision-based in fields of application: gesture recognition for robotic control, sign language recognition, Image processing. Future work leads to an opportunity to design an efficient, compact and feasible device for gesture recognition based on images. Overcoming the issues mentioned will definitely prove a boon to speech- and hearing-impaired people.

References

A. Kumar, PS. Rathore, V. Dutt, *"An IOT Methodology for Reducing Classification error in face Recognition with the Commuted Concept of Conventional Algorithm"* has been published in IJITEE and into the press Volume-8, Issue-11, September 2019 ISSN:2278-3075

A.V. Dehankar and Dr. Sanjeev Jain, "Using Aepi Method for Hand Gesture Recognition in Varying Background and Blurred Images", *International Conference of Electronics, Communication and Aerospace Technology* (ICECA), 2017.

Singh Rathore, P., Kumar, A., & Gracia-Diaz, V. (2020). A Holistic Methodology for Improved RFID Network Lifetime by Advanced Cluster Head Selection using Dragonfly Algorithm. *International Journal of Interactive Multimedia and Artificial Intelligence*, 6 (Regular Issue), 8. http://doi.org/10.9781/ijimai.2020.05.003

Abhishek Jain, Lakshita Jain, Ishaan Sharma, Abhishek Chauhan, "Image Processing Based Speaking System for Mute People Using Hand Gestures", *International Journal of Engineering Sciences & Research Technology* P. If: 5.16, 2018.

Aruljothy. S, Arunkumar. S, Ajitraj. G, Yayad Damodran. D, Jeevanantham. J, Dr. M. Subba, "Hand Gesture Recognition Using Image Processing for Visually Impaired and Dumb Person.", *International Journal of Advanced Research in Computer and Communication Engineering*, 2018.

Bin Feng, Fangzi He, Xinggang Wang, Yongjiang Wu, Hao Wang,Sihua Yi, And Wenyu Liu," Depth-Projection-Map-Based Bag of Contour fragments for Robust Hand Gesture Recognition", *IEEE Transactions on Human-Machine Systems* (Volume: 47 , Issue: 4), 2017.

Dimitry Ryumin, Denis Ivanko, Alexander Axyonov, Ildar Kagirov, Alexy Karpov, Milos Zelezny, "Human – Robot Interaction with Smart Shopping Trolley Using Sign Language: Data Collection", *1st International Workshop on Pervasive Computing and Spoken Dialogue Systems Technology*, 2019.

Du Jiang, Zujia Zheng, Gongfa Li, Ying Sun, Jianyi Kong, Guozhang Jiang, Hegen Xiong, Bo Tao, Shuang Xu, Hui Yu, Honghai Liu, Zhaojie Ju, *Gesture Recognition Based on Binocular Vision*, Springer Science+ Business Media, LLC, 2018.

Guangming Zhu, Liang Zhang, Peiyi Shen, (Member, IEEE), And Juan Song, "Multimodal Gesture Recognition Using 3-D Convolution And Convolutional LSTM", *IEEE Access* (Volume: 5), 2017.

Jiaqing Liu, Kotaro Furusawa, Tomoko Tateyama, Yutaro Iwamoto,Yen-Wei Chen, "An Improved Hand Gesture Recognition with Two-Stage Convolution Neural Networks Using a Hand Color Image and Its Pseudo-Depth Image", *IEEE International Conference on Image Processing (ICIP)*, 2019.

Kamal Preet Kour, Dr. Lini Mathew, "Sign Language Recognition Using Image Processing", *International Journal of Advanced Research in Computer Science and Software Engineering*, Volume-7, Isuue-8, 2017.

Liwei Yang, Yiduo Zhu and Tao Li, "Towards Computer-Aided Sign Language Recognition Technique", *IEEE 4th Advanced Information Technology, Electronic and Automation Control Conference (IAEAC)*, 2019.

Maisyasenand Shaidahjusoh, "A Systematic Review On Hand Gesture Recognition Techniques, Challenges and Applications", *PeerJ Computer Science*, 2019.

Mohamed A. Rady, Sherin M. Youssef, Salema F. Fayed, "Smart Gesture-Based Control in Human Computer Interaction Applications for Special-Need People", *Novel Intelligent and Leading Emerging Sciences Conference (NILES)*, 2019.

N. Bhargava, A. Kumar Sharma, A. Kumar and P. S. Rathore, "An adaptive method for edge preserving denoising," *2017 2nd International Conference on Communication and Electronics Systems (ICCES)*, Coimbatore, 2017, pp. 600–604, doi: 10.1109/CESYS.2017.8321149.

Nagul Nagpal, Dr. Arun Mitra and Dr. Pankaj Agarwal, "Design Issue and Proposed Implementation of Communication Aid for Deaf & Dumb People", *International Journal on Recent and Innovation Trends in Computing and Communication*, 2015.

Oyebade K. Oyedotun, Adnan Khashman, "Deep Learning in Vision-Based Static Hand Gesture Recognition", *Neural Computing & Applications*, 2016.

Pichao Wang, Member, Wanqing Li, Zhimin Gao, Chang Tang, And Philip Ogunbona," Depth Pooling Based Large-Scale 3d Action Recognition With Convolutional Neural Networks", *IEEE Transactions on Multimedia*, 2018.

Polina Yanovich, Carol Neidle, Dimitris Metaxas, "Detection of Major ASL Sign Types in Continuous Signing for ASL Recognition", *Proceedings of the Tenth International Conference on Language Resources and Evaluation (LREC)*, 2016.

Prajwal Paudyal, Junghyo Lee, Azamat Kamzin, Mohamad Soudki, Ayan Banerjee, Sandeep K.S, "Explainable AI for Sign Language Learning", IUI Workshop, 2019.

Rishabh Agarwal, Nikita Gupta, "Real Time Hand Gesture Recognition for Human Computer Interaction", *IEEE 6th International Conference On Advanced Computing*, 2016.

Saeideh Ghanbari Azar And Hadi Seyedarabi, "Trajectory-Based Recognition of Dynamic Persian Sign Language Using Hidden Markov Model", *Elsevier Journal of Computer Speech & Language*, 2019.

Saransh Sharma, Samyak Jain, Khushboo, "A Static Hand Gesture And Face Recognition System For Blind People", *6th International Conference On Signal Processing And Integrated Networks (SPIN)*, 2019.

Shining Song, Dongsong Yan, Yongjun Xie, "Design of Control System Based on Hand Gesture Recognition", *IEEE 15th International Conference on Networking, Sensing and Control (ICNSC)*, 2018.

Rathore, P.S., Chatterjee, J.M., Kumar, A. *et al.* Energy-efficient cluster head selection through relay approach for WSN. *J Supercomput* (2021). https://doi.org/10.1007/s11227-020-03593-4

Shweta S. Shinde, Rajesh M. Autee, Vitthal K. Bhosale, "Real Time Two Way Communication Approach for Hearing Impaired And Dumb Person Based on Image Processing", *IEEE International Conference on Computational Intelligence and Computing Research (ICCIC)*, 2016.

Tong Du, Xuemei Ren, Huichao Li, "Gesture Recognition Method Based on Deep Learning", *33rd Youth Academic Annual Conference of Chinese Association of Automation (YAC)*, 2018.

15

SPAM Filtering Using Artificial Intelligence

Abha Jain

M. Tech Scholar[1], MDS University, Ajmer, India

Abstract

So long as all of us are managing the problem of unsolicited email there may be a dire need for the development of steadfast and strong antispam filters. These days, gadget learning, which a subset of artificial intelligence, is getting used to finding out and determining antispam filters. In this paper we speak of system learning–based electronic mail spam filtering mechanisms and algorithms. It will examine diverse thoughts, attempts, efficiency and different studies of trends in junk mail filtering. The history explains the packages of device gaining knowledge of strategies to clear out the antispam emails of main email service carriers like Gmail, Yahoo, Outlook and so on. We will discuss the unsolicited email filtering techniques and sundry efforts made by various researchers in fighting unsolicited emails via device mastering strategies. Here, we make comparisons of the strengths and weaknesses of already existing machine learning algorithms and techniques and different open studies troubles in spam filtering. We might suggest gaining deep knowledge and also deep learning about adversaries as these technologies will make us capable of efficaciously dealing with the threat of spam emails.

Keywords: Spam filtering, antispam filters, machine learning, artificial intelligence, neural networks, email spamming, clustering

15.1 Introduction

Bulk emails known as unsolicited mail have grown to become a huge nuisance on the internet these days. A spammer is someone who sends these

Email: abhajain001@gmail.com

Neeraj Bhargava, Ritu Bhargava, Pramod Singh Rathore, and Rashmi Agrawal (eds.) Artificial Intelligence and Data Mining Approaches in Security Frameworks, (261–292) © 2021 Scrivener Publishing LLC

illegitimate emails. They acquire the email addresses from diverse websites, social media systems, chatrooms, surveys, contact bureaucracy and so on. Unsolicited mail emails are responsible for wasting of a person's time, garage capacity of the system and community bandwidth. Those huge volumes of junk mail emails which can be coming at the pc community have extreme adverse impacts at the server's memory, communique bandwidth, cpu power and user's time. More than 77% of the worldwide emails are junk mail emails, and these junk mail emails are increasing every year. That is very aggravating for the users who get emails which they have not requested, which are known as junk mail.

The worst thing is that many customers are being trapped by the spammers and hackers who send those junk mail emails to entice the customers into falling for the internet scams and fraudulent practices of the spammers. They pretend to be from trustworthy companies but their intentions are to persuade people to reveal their non-public information like passwords, bank account info, credit score card numbers and so forth [1].

In 2018, a 4th zone survey suggested that spam emails are 51% of the overall emails. Spam fallouts for unproductive use of assets on simple mail switch protocol (SMTP) as they want a method to send a big volume of unrequired messages known as unsolicited mail. In Figure 15.1, we showcased the extent of unsolicited mail messages containing viruses and malware codes in 2018 and 2019 as follows:

Leading Email service companies like Yahoo, Outlook and Gmail have employed diverse combinations of gadgets getting to know techniques like neural networks in its unsolicited mail filters to efficiently deal with the threat posed by email spams. From massive collection computer systems, those machine learning techniques have the capacity to learn and identify spam emails and unrequired messages by way of analysing comparable messages. The use of pre-current regulations, gadget learning techniques adapt to various conditions and perform more than just checking the Gmail and Yahoo junk emails. These device learning fashions make new standards themselves depending on what they have learned throughout these junk mail filtering operations.

Google has the most superior machine studying model that could hit upon and filter out junk mail and phishing with 99% precision. The result of this is that only one out of a thousand messages will be successful with regard to escaping the junk mail channel. In line with Google statistics,

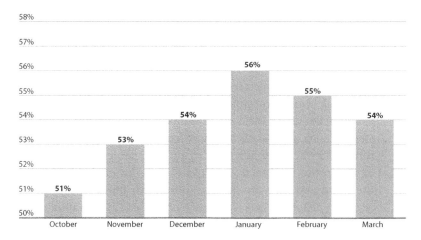

Figure 15.1 The capacity of spam emails 4th zone 2018 to 1st area 2019.

50 to 70% of emails that can be received by way of Gmail are spam mails [2].

Google tools like secure surfing for classifying web sites which have malicious URLs are integrated through Google detection fashions. The overall performance of Google's phishing detection is made more effective by means of the introduction of a device that suspends the shipping of emails and messages for some time to hold additional extreme inspection of the phishing messages as they're comparatively smooth to stumble on while they're analyzed in bulk. These phishing messages and junk mail emails are purposely introduced late to conduct the deeper exam of these suspicious emails at the same time as other messages are introduced on time and the algorithms are updated in real time. The emails which can be treated by this postponement amount to only 0.05%.

Below are the special categories of junk mail filtering techniques which are widely used to resolve the problem of unsolicited mail emails:

1. Content Material-Based Totally Junk Mail Filtering Approach: This approach analyses phrases, their occurrences, and the distribution of words and phrases in the frame of an email and then uses them to filter the incoming unsolicited emails. This is used to create automatic filtering guidelines and to categorize emails by the use of system

mastering algorithms like Naïve Bayes classification, SVM, ok-nearest neighbour, and neural network. Those algorithms will be mentioned in a later section.

2. Case-Based Junk Mail Filtering Method: That is the maximum popular spam filtering strategies in which each junk mail and non-spam emails are pulled out from a person's email by the use of the collection model. Then pre-processing steps are accomplished to transform the emails using customer interface, characteristic extraction, choice, a grouping of electronic mail facts, and examining the method. Then the examined facts are divided into vector units. Sooner or later, the machine gets to know techniques that are used to educate datasets and tests the records to decide whether or not the incoming electronic mail is a junk mail or not.

3. Rule-Based or Heuristic Junk Mail Filtering Approach: This method checks a big range of styles which might be ordinary expressions against chosen messages the usage of present guidelines. The rating of the messages increases with those similar patterns, or it removes for the rating if any styles did not suit. The messages might be considered as junk mail, if the message exceeds the unique threshold, in any other case may be taken into consideration as non-unsolicited mail. Those techniques need to be updated through the years, which is a good way to cope successfully with spammers who are constantly introducing new junk mail messages that can without difficulty escape being recognized as unsolicited mail. An example of a rule-based junk mail clear out is a spam slayer.

4. Past Likeness–Based Spam Filtering Approach: This approach utilizes memory-based, or instance primarily based, AI techniques to sort incoming messages dependent on their closeness to put away messages. The houses of the email are utilized to make a multi-dimensional space vector, which is applied to plot new cases as focuses. These new cases are a quick time later allocated to the most a way accomplishing class of its okay-nearest preparing cases. This technique utilizes the k-closest neighbour (KNN) for cleaning spam messages.

5. Adaptive Unsolicited Mail Filtering Technique: This approach senses and filters unsolicited mail by grouping them into distinct lessons. It divides an email frame into diverse corporations; each organization has a symbolic textual content.

An assessment is made between every incoming e-mail and each organization, and then a percentage of resemblance is calculated to pick out the possible institution the email belongs to.

15.2 Architecture of Email Servers and Email Processing Stages

15.2.1 Architecture - Email Spam Filtering

E-mail junk mail filtering refers to reducing unsolicited emails to a bare minimum. In electronic mail filtering, emails are processed and rearranged in accordance with a few positive requirements. Email filters are used to control the following, including incoming emails, clear out junk mail emails, come across and take away emails that encompass malicious codes which include viruses and malware. SMTP protocol is answerable for the running of emails.

Extraordinary internet provider vendors arrange the spam channels at each layer of the framework, earlier than the email expert, or at mail circulate wherein there's the proximity of a firewall. Primarily based at the decided protection policies, the firewall, which is a network security system, manages and video display units the incoming and outgoing network visitors. The email server aids in implementing anti-unsolicited mail and an anti-virus solution enable in presenting complete safety measures for email at the circumference of the community. Add-ons are set up between the endpoint gadgets to function as a middleman in which the filters may be carried out in clients [3].

How Gmail, Yahoo and Outlook Emails Spam Filters Work
Gmail, Outlook.com and Yahoo actively operate on various spam filtering formulas to deliver the most effective authorized emails to the users and clear out legitimate or junk emails. On the other hand, those formulations occasionally mistakenly block dependable messages. The report says, approximately 23% of the relevant emails commonly fail to reach the inbox of the actual recipient. Various mechanisms are designed through the email service providers to use anti-spam filters to limit the threats modelled by the means of malware, phishing and many others to email users. Many mechanisms are used to determine the danger level for each incoming email. These include first-rate spam limits, sender policy frameworks,

whitelists and blacklists, and recipient verification equipment. Those mechanisms are designed to be utilized by single or a couple of users.

15.2.1.1   Spam Filter - Gmail

Hundreds of rules are used by Google records centers to assess the authenticity of an electronic mail. Depending on the chance of whether or not the feature is spam or not, every rule describes the features of an email to be junk mail and the sure statistical fee is attached to it. Then with the use of this weighted importance, constructions of equations are performed. The rating is used in opposition to a sensitivity threshold to conduct a test decided by means of the user's junk mail clear out. Hence, it is labeled as unsolicited mail email or valid mail.

For the class of emails, Google uses its spam recognition gadget learning algorithms like logistic regression and neural networks. Google additionally practices Optical Individual Popularity (OCR) to guard its customers' shape picture spam. Device learning algorithms are designed and developed to mix and rank the big sets of search engine search consequences as a result of which, a customer is allowed to link various factors of Gmail to simplify the junk mail category [4].

Area popularity and link headers are the essential elements responsible for the evolution of unsolicited mail over time, due to these messages are suddenly become in the junk mail folder.

15.2.1.2   Mail Filter Spam - Yahoo

Yahoo is the largest unrestricted webmail specialist co-op with 320 million clients worldwide. It has its own calculations that are utilized to identify spam messages. URL Filtering, email substance and spam objection from clients are the fundamental techniques utilized by Yahoo to recognize spam messages.

Yahoo filters the emails by way of an area not unlike Gmail, which filters the emails by using IP cope with. Yahoo has its own mechanisms to prevent legitimate customers from being incorrectly identified as a spammer. Unlike blacklisting, Yahoo gives internal whitelisting and return certification with the assistance of letting the users specify the list of receivers and senders.

These spam filters allow the person to use a mixture of whitelist and other unsolicited mail-combating features as a way of lessening the variety of valid messages which can be erroneously categorized as unsolicited mail. Then again, the use of whitelist will make the filter out to be very strict and the implication is that any unapproved consumer would be blocked automatically. Many anti-spam systems use automated whitelist. In this example, a nameless sender's electronic mail is checked in opposition to the database; if there are no spamming records, their message is delivered to the receiver's inbox [5].

15.2.1.3   Email Spam Filter - Outlook

Outlook.com changed into Microsoft's metro plan language and directly duplicates the interface of Microsoft Outlook.

The email header is included fields, for example, the sender's region, the beneficiary's location, or timestamp which show whilst the message changed into sent through transitional workers to the message transport dealers that act as an office for arranging sends. The header line, for the most component, begins with a "from" and it experiences a few adjustments at anything factor it movements starting with one employee then onto the following through an in the middle worker. Headers permit the consumer to look at the route the email goes through, and the time taken by way of each employee to treat the mail. The reachable facts need to go through some coping with earlier than the classifier can put it to use for separating.

15.2.2   Email Spam Filtering - Process

Email Messages are produced from massive components that are the header and the frame. The header is the territory that has expansive statistics about the substance of the email. It includes the problem, sender, and beneficiary. The frame is the middle of the email. It could contain information that doesn't have a pre-characterized data.

Fashions comprise a website page, sound, video, simple information, pictures, documents, and HTML markup. The email header is protected fields, as an example, the sender's location, the beneficiary's region, or timestamp which display while the message became sent by transitional people to the message transport sellers that acts as a workplace for arranging sends. The header line for the maximum part begins with a "from" and it studies a few adjustments at something point it movements starting with one worker then onto the next via a within the center worker. Headers permit the consumer to see the course the email goes through, and the time taken through every worker to deal with the mail. The reachable statistics wishes to undergo a few coping with earlier than the classifier can put it to use for isolating shown in Figure 15.2 [6].

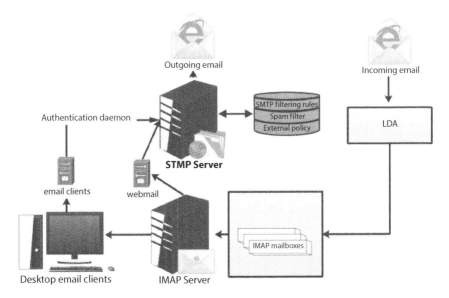

Figure 15.2 Representation of an email server structure and process of spam filtering.

15.2.2.1   Pre-Handling

This is the foremost degree this is completed at something point a drawing near mail is gotten. This development contains tokenization.

15.2.2.2   Taxation

This method expels the words within the frame of an electronic mail. It additionally changes a message to its enormous elements. It takes the email and partitions it into an arrangement of delegate pictures called tokens. These delegate snapshots are extricated from the frame of the email, the header, and concern. Some scientists accentuated that the way closer to supplanting information with precise distinguishing evidence snapshots will dispose of all of the qualities and phrases from the email restrictive of considering the significance.

15.2.2.3   Election of Features

Sequel to the pre-making ready level is the element choice stage. Highlight preference a type of lower in the proportion of spatial inclusion that viably famous fascinating elements of the email message as a packed element vector. The technique is gainful while the size of the message is tremendous and a summed up portrayal is predicted to make the challenge of textual content or photograph coordinating smart. Improve price extortion, inclusive of legacy, lottery, visa and customs-leeway hints, romance tricks, which include selling medicines to improve sexual performance, internet dating, army hints, commercials for pornography, commercials for incidental out-of-doors destinations, promising big money "telecommute" occupations, primarily web-based shopping, strategic agreements and others.

Likely the maximum valued highlights for spam filtering consist of: Message Frame and the extent of the message, occurrence take a look at of words, daily examples of messages (unsolicited emails most of the time have numerous semantic inconsistencies), recipient age, sex and nationality, recipient responded (shows whether the beneficiary answered the message), adult substance and bag of phrases from the message content [7].

Sender account capabilities applied for unsolicited mail isolating include sender USA (the move of nations as expressed through customers on their profile and as uncovered through their IP address), sender IP cope with, sender electronic mail, sender and recipient age, sender popularity. The less extensive highlights are geographical separation amongst sender and beneficiary, sender's date of beginning, username and mystery phrase of the sender, account lifestyles expectancy, gender of sender, and age of the beneficiary. The acknowledgment of junk mail messages with the least range of highlights is good-sized considering computational multifaceted nature and time. Spotlight choice consists of paperwork like stemming, commotion expulsion, and stop word evacuation steps.

15.2.3   Freely Available Email Spam Collection

The dataset contained in a group assumes an enormous process in surveying the exhibition of any spam channel. Despite the truth that there are numerous customary datasets which are generally applied for grouping textual content, a few scientists within the area of junk mail sifting are making the gathering applied for assessing the adequacy of their proposed channel reachable to human beings in the standard.

15.3   Execution Evaluation Measures

Generally, Spam networks are evaluated on huge databases pertaining to ham and unsolicited emails which can be overtly available to clients. It is a case of execution quantifies which might be applied is type accuracy (acc).

It is the relatively wide variety of messages appropriately ordered, the level of messages nicely characterized is utilized as an extra measure for assessing the execution of the channel. It has anyway been featured that utilizing accuracy as the main execution lists isn't good enough.

Different execution measurements, for example, overview, accuracy, and inferred measures utilized inside the field of information restoration ought to be an idea of, so additionally bogus positives and bogus negatives are applied in preference hypothesis. This is considerable due to the costs related to misclassification. At the factor, while an unsolicited email message is wrongly named ham, it offers to ascend to a few degrees beside the point difficulty, in spite of the fact that the main thing the customer needs to do is to erase such a message. Conversely, when a non-unsolicited email message is improperly marked as spam, this demonstrates the hazard of dropping considerable information because of the channel's grouping mistake.

This is fundamental especially in settings where unsolicited mail messages are therefore erased. In this way, it is inadequate to assess the exhibition of any machine gaining knowledge of calculation applied in junk mail channels utilizing grouping precision totally. Besides, in a setting this is disproportionate or one-sided in which the amount of spam messages used for trying out the presentation of the channel is a lot better than that of ham messages, the classifier can report a high precision through focusing on the discovery of junk mail messages completely shown in Table 15.1 [8].

In a situation, in which there isn't always zero probability of wrongly classifying a "not a spam" or ham message, it's a necessity that a change-off be reached among the two types of errors, contingent upon the inclination of consumer and the exhibition markers applied. The formulae for calculating the class accuracy and class mistakes are depicted in Eq. (15.1) and (15.2) under [9]:

Let us consider:

N_h= Quantity of non-junk emails to be classified.

N_s= Range of unsolicited emails to be categorized

$$\text{Classification Acc (Acc)} = |h \rightarrow h| + |s \rightarrow s| \qquad (15.1)$$

*Acc = Accuracy

Table 15.1 Freely available email spam collection.

Name of dataset	Messages count		Spam rate	Year
	Spam	**Non spam**		
Backlog Spam	1590	0	100%	1998
Spam-Base	1813	2788	39%	1999
Ling-Spam	481	3412	17%	2000
PU1	481	618	44%	2000
PU2	1897	4150	31%	2002
PU3	142	579	20%	2003
ZH1	1826	2313	44%	2003
Gen-Spam	571	571	50%	2003
Trec 2005	1205	428	74%	2004
Biggo	31,196	9212	78%	2005
Phishing Corpus	52,790	39,399	57%	2005
Trec 2006	8549	0	100%	2005
Trec 2007	415	0	100%	2010

$$\text{Classification Error (Err)}= 1-\text{Acc}=\frac{|h{\to}s|+|s{\to}h|}{N_h+N_s} \qquad (15.2)$$

Arrangement Accuracy and Error normally consider distorted absolute $|h{\to}s|$ and distorted minus $|s{\to}h|$ presences to endure equivalent expense. It is fundamental to call attention to it that unbalanced blunder costs are engaged with spam sifting. Wrongly grouping a spam (otherwise called bogus positive occasion) is a costly blunder contrasted with the spam message simply sidestepping the channel. Such occurrence is referred to as bogus negative occasion. At the point when a genuine email is properly named ham, it is known as a genuine positive occasion $|h{\to}h|$. By the by, when a spam email is appropriately labeled as spam, at that point a genuine negative occasion |s→s| has occurred. In view of the above clarifications, the Distorted Absolute Rate (DAR) is characterized as the proportion of legitimate emails that are delegated spam. It is meant utilizing the equation in Eq. (15.3) below

$$DAR = \frac{No.\ of\ Distorted\ Absolute}{No.\ of\ Distorted\ Absolute + No.\ of\ True\ Negatives} \qquad (15.3)$$

Additionally, permitting unsolicited emails which have been inflamed with malware, adware, spyware, Trojan, botnet, viruses, worms, or phishing baits consisting of messages claiming to be from Social websites, dating websites, auction websites, banks, online payment processors are usually used to entrap the recipients. This leaves the customer open to huge losses. The ratio of spam messages that are wrongly labelled as legitimate is known as the Distorted Negative Rate (DNR). That one is extra apt metric for comparing the performance of a filter out. The formula for computing the fair is in eq. (15.4) below, allowing spam emails which have been inflamed with malware includes messages claiming to be from social websites, dating sites, auction websites, banks, online payment processors are generally used to entrap victims [10, 11].

The ratio of unsolicited emails that are wrongly labelled as legitimate is called Distorted-Negative Charge (DNC). That is an extra apt metric for evaluating the overall performance of a filter. The system for computing the DNR is in eq. (15.4).

$$DNR = \frac{No.\ of\ Distorted\ Negatives}{No.\ of\ True\ absolutes + No.\ of\ Distorted\ Negatives} \qquad (15.4)$$

Spam filters with low DAR and DNR have a good performance. These characteristics (DNR and DAR) represent the performance of filters without delay goal on the classification selection borderline without producing the probability estimate. The filters efficiency that estimates the organization conditional chances after which it perform classification is based totally on estimated possibilities which can be represented through a curve which is known as ROC (Receiver Operating Characteristic) Curve. ROC Curve is a graphical plot that exhibits the analytical functionality of a junk email filter as its bias level is changed. The ROC Curve is generated by using plotting the real nice rate against the false wonderful price (DAR) at one threshold setting. The actual high-quality rate is referred to as Sensitivity. The false-tremendous price is called the probability of fake alarm which is computed with the aid of subtracting the fee of the specificity from 1 (i.e. E. 1 - specificity).

Two imported metrics from the sector of statistics recovery 'remember' and 'precision' are used to acquire the effectiveness and function of spam filters. For the reason where:

$$|S \to Ns| = \text{Counts of spam emails categorized as non-spam}$$

$$|Ns \to S| = \text{Counts of non-spam emails named spam separately}$$

And here $|Ns \to Ns|$ and $|Ns \to S|$

Eq. (15.5) below represents spam 'remembers' (R_s) and spam precision (Ps):

$$R_s = \frac{|S \to S|}{|S \to S| + |S \to H|} \text{ and } P_s = \frac{|S \to S|}{|S \to S| + |H \to S|} \tag{15.5}$$

Remember (R_s) also known as effectiveness can be defined as the relatively wide variety of spam messages that filter out succeeded in preventing from coming an email in inbox. Additionally, Precision (Ps) described because the reliability of the filter out is calculated by means of dividing the range of messages classified with the aid of the filter out as spam but are genuinely legitimate by using the full variety of emails. Evaluating the performances of different junk mail filters, the use of (Rs) and (Ps) is sensitive thinking about the extraordinary value that have concerned within the computations that produced (Rs) and (Ps). The value of fake positives is (lambda λ times) than that of the negatives, where (λ) is a component, numerical in nature, which specifies how 'volatile is to classify a legitimate email as a junk email. Remuneration understanding must be taken into account which can be done by making each of the legitimate emails as equal to α email [12].

In this liquidation, components used for computing price related measures consisting of Complete Cost Proportion (CCP) and Weighted Accuracy (WAcc) is discussed.

The complete cost proportion is used for measuring the accuracy of filters. Higher CCP shows better overall performance. Whilst the fee of CCP <1, it's far higher no longer to use the clear out. In a situation where the price is balanced to time squandered, CCP measures the quantity of time squandered by the user to delete all unsolicited email by regardless of the fact that junk email filter is set up. It then compares it to the time spent to manually dispose the unsolicited email that stay away from the filter out in addition to the time required to recoup from legitimate messages that had been mistakenly blocked.

The two main strengths of CCP are that it is by far an unmarried-parent dimension, whilst the majority of the alternative cost touchy measures require at the very least two figures. This though can provide the incorrect effect about the effectiveness of a filter as a higher CCP may denote a significantly decreased rate or a completely high. Likewise, CCP is prone to the stabilizing of gathering. The stableness of gathering is a situation where the extent of unsolicited mail and no spam messages inside the series are at variance. The portability of the values is one of the drawbacks of the CCP. Also, contrast can be best drawn among CCP values while all evaluated CCPs are calculated by the means of comparable λ. The computing formular for Weighted Accuracy (WAcc), Weighted Mistakes Fee (Wherr), and Entire Fee Share (CCP) are represented in EQs. (15.6), (15.7), and (15.8) underneath:

$$W_{Acc} = \frac{\lambda |H \rightarrow H| + |S \rightarrow S|}{N_H + N_S} \text{ and } W_{Acc} = 1 - W_{Acc} \tag{15.6}$$

$$W_{Acc} = \frac{\lambda |H \rightarrow S| + |S \rightarrow H|}{N_H + N_S} \tag{15.7}$$

$$TCR = \frac{N_S}{\lambda |H \rightarrow S| + |S \rightarrow H|} \tag{15.8}$$

While calculating sensitivity of filters, lambda (λ) comes to a decision for wrongly classifying a non-spam email as junk email. Included into the brim is the sensitivity value with the components λ/(1 + λ). The version is rebuilt and is measured on various styles of firmness degree of λ.

The accurateness measure of a test is described as the weighted harmonic suggested of the precision (Ps) and don't forget (Rs) of the test in an equation. F-measure uses a parameter that permits negotiation to be reached about recall and precision. F1 represents conventional f-degree which is normally used and gives uniform weight to the memory and precision as shown in Eqs. (15.9) and (15.10).

$$F_1 = \frac{2*recall*precision}{Precision + Recall} \tag{15.9}$$

$$F_{Beta} = \frac{(1 + Bet\ a_2)*\text{recall}*\text{precision}}{\text{Recall} + Bet\ a_2*\text{Precision} + \text{Recall}} \qquad (15.10)$$

In a state of affairs wherein we've 0 < beta<1, provides more precision while, when we have beta>1, it gives provides greater significance. It is superficial that f-measure is a special case of weighted f-degree when beta = 1 [13, 14].

15.4   Classification - Machine Learning Technique for Email Spam

Of later, unsolicited mail characterization is in the main treated through AI (ML) calculations deliberate to split among spam and non-unsolicited mail messages. AI calculations accomplish this via a programmed and versatile approach. As opposed to depending upon hand-coded makes a decision that is helpless in the face of the interminably changing traits of junk mail messages, ML strategies have the capacity to get statistics from a variety of messages given, and later on, make use of the procured statistics to group new messages that it actually got. Ml calculations have the capability to carry out higher depending on their revel in. In this phase, we will survey likely the maximum well-known AI strategies which have been carried out in the junk mail area.

15.4.1   Flock Technique - Clustering

Clustering is organizing a set of patterns into related training. Clustering technique is used in dividing case investigations into moderately similar groups called clusters. Clustering strategies have involved many researchers and academicians in different fields of application.

Clustering algorithms that can be unsupervised learning gear are used on junk email datasets which typically have genuine labels.

Two kinds of clustering methods that are used for junk emails are density-primarily based clustering and k-nearest neighbour (KNN). Density-based is a clustering method in which any other approach of report that has been exploited to clear up the spam. This method has the potential to encrypt messages, which in turn upholds the confidentiality of the message.

KNN is a distribution-based method, that no longer relies on the assumptions that the information is drawn from a given probability distribution. Nearly all the applied statistics disobey the usual hypothetical postulations made (which includes gaussian combination, linearly separable, and others). Non-parametric algorithms like KNN can be used to save this kind of scenario. In the KNN classifier, the classification model isn't always built from statistics, instead, the category is completed through matching the test example with various training datasets, and a choice is made as to which group it belongs to relying on the resemblance to training datasets closest associates [15].

The KNN is also known as a lazy learner as the training datasets are not used to perform generalization.

Various KNN set of rules for filtering spam mails are explained in the algorithm underneath. Here neighbours (r) return the k nearest friends of r, closest (r, t) go back to the nearest factors of **t** in **r**, and take a look at class(D) go back to the elegance label **s**. An easy KNN set of rules for unsolicited email type is inside the set of rules below:

Algorithm: Classification of Spam Email
Step 1. In first of this algorithm find email messages with labels.
Step 2. **k** is no. of closest neighbours.
Step 3. **E** shows Email Test Messages.
Step 4. **T** indicates the set of Training Email Message.
Step 5. Labelled as set of Email Messages is **L**.
Step 6. Interpret training data files.
Step 7. Interpret the testing data files.
Step 8. for each r in E and each t in T do
Step 9. Neighbours (r) = {}
Step 10. If| Neighbours(r)| <k then
Step 11. Neighbours (r) = Closest (r, t) U Neighbours(r)
Step 12. End if.
Step 13. If| Neighbours(r)| >k then
Step 14. Contain (M,x_j,y_j)
Step 15. end if.
Step 16. end for 17: return Final Email Message Classification (Spam/Valid email).
Step 17. End.

15.4.2   Naïve Bayes Classifier

Naïve Bayes' Classification Algorithm is named after Thomas Bayes (1702–1761), who gave the algorithm. Bayesian Classifier demonstrates a supervised mastering technique and a statistical technique for type. It acts as a probabilistic model that facilitates to grab doubts approximately the model in an ethical manner by means of manipulating the probabilities of the consequences. It is used to provide a method for analytical and predictive troubles. The category gives sensible gaining knowledge of algorithms of preceding expertise and experimental facts that may be merged. This type gives a beneficial view of knowledge and comparing numerous getting to know algorithms. It computes precise probabilities for principle and it's miles strong to noise in input facts.

Naive Bayes classifier is a forthright probabilistic classifier that is founded on Bayes theorem with assumptions that can be impartial in nature. A better expression for the opportunity version has to be the self-reliant characteristic version is shown in eq. (15.11):

$$\text{Theorem: } P (B \text{ given } A) = P (A \text{ and } B)/P (A) \qquad (15.11)$$

Concept of complexity preventive sovereignty is made to make calculation simpler, and this is the idea of naming the algorithm as 'naïve'. However, the set of rules is powerful and really sturdy. It plays just like other supervised algorithms. There is a growth inside the reputation of NB as a simple and computationally efficient algorithm with high-quality performances in solving actual-world problems. As a result of its exquisite qualities, NB classifiers have found software as a category algorithm in textual content, junk mail email, sentiment evaluation, recommender systems, junk mail opinions, and different online packages.

Naïve Bayes classifiers are particularly utilized in textual content classification (as it produces an advanced result in multi-class troubles and independence rule) and have an extra fulfilment price while compared to a few different gadget learning algorithms. Because of this obvious gain, its miles significantly applied inside the subject of unsolicited mail filtering (locate unsolicited mail email) and sentiment evaluation (in social media evaluation, to understand advantageous and negative client evaluations). Spam filtering is the most famous use of the NB classifier. This is a popular method for distinguishing unsolicited mail emails from the authenticate emails, called ham [16].

Most mail customers apply Bayesian spam sifting calculation. Basically all the measurement-based spam separating procedures are utilizing Naïve Bayes' classifier to bunch the insights of every token to an absolute score and the score is utilized in making goal on the sifting. The symbolic T which means the spam rating is processed as outlined in Eq. (15.12):

$$S[T] = \frac{C_{spam}(T)}{C_{spam}(T) + C_{Ham}(T)} \qquad (15.12)$$

Where

$C_{Spam}(T)$ = The number of spam messages containing token T,

$C_{Ham}(T)$ = The number of ham messages containing token T,

There is a need to join the distinctive spamming of the tokens to calculate the spaminess of message to compute the possibility of a message **m** with tokens **t1,......,tn**. Calculating the particular token's spaminess and comparing it with the manufactured from a specific token's hamminess is a straightforward way to make classifications. It is represented in eq. (15.13) below:

$$\left(H[M] = \prod_{i=1}^{N} (1 - S[T_i]) \right) \tag{15.13}$$

The message is delegated spam if the all out maliciousness item S [M] is more noteworthy than the hamminess item H [M]. The above portrayal is utilized in Bayes' grouping calculation for spam email order portrayed underneath:

Naïve Bayes' Category Algorithm for Email Unsolicited Type

Step 1. Enter email dataset

Step 2. Parse every email into its issue tokens.

Step 3. Compute chance for each token S [w] = $C_{spam}(W)/(C_{ham}(W) + C_{spam}(W))$

Step 4. Database Stores the value of spaminess

Step 5. for each **M** message do

Step 6. while (M does not end) do

Step 7. Examine email for the following symbolic Ti.

Step 8. Question the database for spamminess S(Ti).

Step 9. Calculate possibilities of emails accrued S [M] and H [M].

Step 10. Figure the complete email sifting signal by: I [M] = f (S [M], H [M])

Step 11. I[M]= I+S[M]−H[M]/2

Step 12. if I [M] > threshold then

Step 13. email marked as spam

Step 14. else

Step 15. email marked as non-spam.

Step 16. end-if

Step 17. end while

Step 18. end for 19: return Final Email Classification (Spam/Valid email)

Step 19. End.

15.4.3   Neural Network

Neural Networks are gatherings of sincere handling gadgets that are interconnected and talk with every different by means of strategies for a giant range of weighted associations. Every one of the gadgets acknowledges contribution from the neighbouring gadgets and outdoor resources and figures the yield this is communicated to different acquaintances. The mechanism for calibrating masses of the associations is likewise made reachable. Neural networks are an effective calculation for taking care of any AI issue that calls for characterization. Due to their creativity, they're advancing as a giant device in the AI scientist's arrangement of apparatuses. Although, neural structures aren't typically utilized inside the region of junk mail email as one may additionally conceivably believe.

Naïve Bayes is an incredible method for spam characterization with excessive precision (99%) and a false-fine fee. What improvements its high exactness is the giant number of all-round interconnected making ready components (neurons) which can be working in settlement to provide solutions for specific problems. For instance, Google described the growth in Gmail spam channels' precision from 5% to 99.9% within the wake of fusing neural structures into it. This infers neural structures can be useful for improving the exhibition of spam channels, especially when hybridized with Bayesian order and one of kind methods. On the other hand, a good deal of exam has to be completed on the use of neural systems for unsolicited mail discovery, and almost everything of the ebb and waft research takes the system layout, power, and studying rate to be constant. Extra exploration endeavours need to be centred around the adequacy of the gadget throughout datasets rather than the propriety of different system plans for the interest. As indicated through, normally there are three types of units.

- Input Unit: accepts sign from an outside supply.
- Output Unit: transmits data out of doors to the community.
- Hidden Unit: accepts and transmits alerts within the network

The operations of the framework are synchronized so an extraordinary number of units can work in equal. ANN is altered to get a lot of information sources and produce the required arrangement of yields. This cycle is known as learning or preparing. There are two kinds of preparing in neural organizations (NN).

- Supervised: organization is prepared by giving the organization a lot of information sources and coordinating yield designs, known as the preparation dataset.
- Unsupervised: organization self-trains by gathering examples.

Neural Networks are of two types which is typically implied whenever Artificial Neural Network is used. They are the perceptron and multilayer perception.

This segment will explain the perceptron algorithm and its software to junk mail filtering. Beneath is a perceptron set of rules which is a well-known neural network set of rules. The perceptron assists in finding a linear feature of the characteristic vector $\mathbf{f(x)} = \mathbf{wTx} + \mathbf{b}$ such that $\mathbf{f(x)} > \mathbf{0}$ for vectors of one organization, and $\mathbf{f(x)} < \mathbf{0}$ for vectors of different group.

Moreover, w = (w1, w2,…wm) are the weights of the characteristic, where $\mathbf{b}$ is the intended bias. The groups may be given the numbers from +1 and -1, so that search for a function is performed. The perceptron learning begins by means of arbitrarily choosing parameters (w0, b0) of the $\mathbf{d(x)} = \mathbf{sign}\ (\mathbf{wTx} + \mathbf{b})$ resolution and time and again bringing them up-to-date.

The training dataset (x, c) is chosen at the $\mathbf{n^{th}}$ repetition of the algorithm to the volume that the present choice characteristic now corporations it as wrong (i.e. Sign (wnx + bn) $\neq$ c). The rule of thumb depicted by eq. (15.14) beneath is utilized in updating the parameters (wn, bn):

$$wn+1 = wn + cx \qquad bn+1 = bn + c \qquad (15.14)$$

The standards for ending the calculation is that a choice capacity must be found which precisely classifies all the preparation tests into various gatherings. There are times when the preparation information can't be isolated directly, in such cases the most astute move to make is to end the preparation calculation once the quantity of information that are incorrectly arranged is adequately little. The algorithm underneath speaks to the calculation for a Perceptron Neural System for email spam arrangement [17]:

Perceptron Neural Community Set of Rules for Email Junk Mail Classification
Step 1: Enter sample electronic mail message dataset
Step 2: Initialize w and b (to random values or to zero).
Step 3: discover a schooling sample of messages (x,c) for which sign ($w^{T}x$+ b).
Step 4 if there is no such pattern, then
Step 5: training is finished
Step 6: keep the final w and stop.

Step 7: else
Step 8: replace (w,b): w= w +cx,
Step 9: b= b+ c
Step 10: go to step eight
Step 11: end if
Step 12: determine electronic mail message elegance assign (w^{T}x+b)
Step 13: go back very last e-mail message type (junk mail/non-unsolicited mail email)
Step 14: end.

The structure of the Neural Network email junk mail classifier is depicted in Figure 15.3 below.

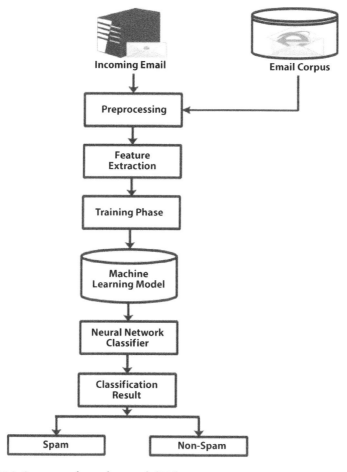

Figure 15.3 Structure of neural network (NN).

15.4.4   Firefly Algorithm

The firefly set of rules (FA) is a populace primarily based met heuristic calculation. It was based on the shimmering behaviour of fireflies. The calculations jam and increment a few up-and-comer arrangements with the aid of methods for population physiognomies to coordinate the inquiry. The structure of the calculation turned into established at the investigation of the idea of correspondence among fireflies at the time they're making ready to have sexual members of the family and fast they're provided to hazard. Fireflies percentage information amongst themselves by using techniques for their shining feature. With round 2,000 firefly species on earth, all use an extraordinary shining configuration. The fireflies commonly create a bit sparkle with a specific corporation problem to what they're engaged with. The mild is produced by means of the biochemical advent of mild by using dwelling animals. Contingent upon the kind of the mild, the proper buddy will impart therefore via either mimicking a similar structure or replying back through utilizing a genuine structure. Alternately, the strength of light decreases inferable from separation. Consequently, a shimmering light radiating from a firefly gets a reaction from fireflies round it inner a visible scope of the blaze. The residences of fascination and improvement of fireflies may want to pass a streamlining calculation where preparations comply with better (extra high-quality) preparations. The firefly calculation for email unsolicited mail arrangement is established as follows:

Algorithm 4. Firefly set of rules for the email spam classification
Step 1. Input electronic mail series with m wide variety of capabilities
Step 2. Set the value of k as 0, i.e k = 0
Step 3. Get population of firefly n
Step 4. Get the quantity of attributes m
Step 5. Set the firefly population
Step 6. For each firefly
Step 7. Choose the firefly which has great health
Step 8. Pick equivalent features from the checking out a part of the e-mail spam collection
Step 9. Take a look at the email message
Step 10. k = k+1
Step 11. Update every firefly
Step 12. Categorize the e-mail message as both spam and non-unsolicited mail electronic mail
Step 13. end for

Step 14. Go back very last e-mail message category (junk mail/non-junk mail e-mail)

Step 15. End

15.4.5   Fuzzy Set Classifiers

The fuzzy set concept was proposed in 1982 out of a push to introduce the correct structure for the robotized transformation of statistics into knowledge. The process is focused around the breakdown of the categorization of anticipated, questionable, or incomplete information expressed as a long way as the records got for a reality. The fuzzy set idea may be depicted as an ongoing numerical strategy to fluffiness. A fuzzy set is based totally on the concept that some facts are associated with every item of the universe. Rs is a mathematical tool that makes a specialty of uncertainty. It's far as in keeping with the thought that any faulty version can be evaluated from below and from overhead by way of utilizing an association that is subtle in nature. One of the enormous additives of the rs concept is the need to discover repetition and conditions among highlights. The fuzzy set idea has been carried out to junk mail sifting since it offers productive and less tedious calculations to extricate hid examples in the facts. It additionally has the ability to relate to facilitating the connections that different regular actual strategies are coming across hard to music down. Except, it acknowledges the usage of both quantitative and subjective information. It has the ability to estimate the minimum sets of data wished for grouping jobs.

Discovering the significance of statistics and growing a set of decision rules from the given facts set is part of the strength of the rs classifiers. It is important to observe that the bushy set idea expresses imprecision by way of the usage of a borderline phase of a fixed in place by using manner of membership. Having the marginal place of a fixed void infers that the set has been plainly characterized (actual) if no longer the set is supposed to be fuzzy (inexact). For a marginal vicinity that consists of on any occasion, one factor within the set connotes that what we consider the set is not enough to exactly depict the set. It tends to be visible that the fuzzy set theory allows clients to evaluate the noteworthiness of records. It allows the consumer to evidently create the arrangements of desire requirements from the statistics. It is simple. It gives a right away translation of were given consequences. It is appropriate for simultaneous (same/disseminated) managing. Figure 15.4 below shows the email separating process work process of the fuzzy set theory from the customer publish the container.

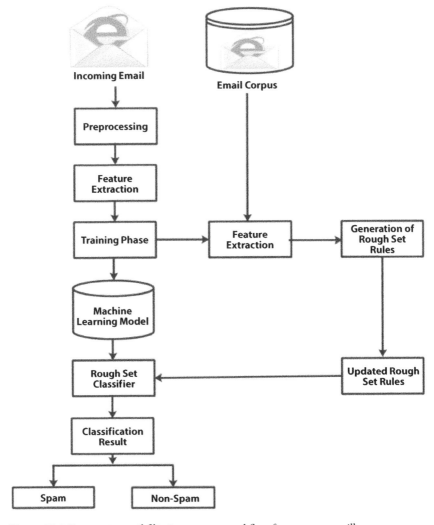

Figure 15.4 Fuzzy set e-mail filtering manner workflow from person mailbox.

15.4.6   Support Vector Machine

Support Vector Machines (SVM) is supervised gaining knowledge of algorithms that perform higher than a few different related studying algorithms. SVM application in presenting solutions to quadratic programming issues that have inequality constraints and linear equality through differentiating one-of-a-kind companies through a hyper plane.

It takes complete benefit of the boundary. Although the SVM might not be as speedy as different type techniques, the set of rules attracts its strength from its high accuracy because of its capability to model multi-dimensional borderlines that are not sequential or truthful. SVM is not without problems prone to a scenario in which a version is unreasonably complicated along with having numerous parameters comparative to the range of observations.

These qualities make SVM the ideal algorithm for software in the regions like:

    a) Virtual handwriting recognition
    b) Text categorization
    c) Speaker popularity, and so on.

Now, c denotes the price parameter to regulate displaying errors that take place while a function is carefully in shape to a restricted set of records points through fining the mistake ξ. In the course of schooling, we presume to have a fixed of statistics to learn, theoretically, there may be most effective a mix of the parameter (c, γ) which has the capacity to provide the superior SVM classifier. On parameters, c and γ grid search is the only possible approach this is implemented in SVM education to acquire the merger of the parameter. The k-fold rotation estimation is energetic inside the grid seeks to pick the SVM classifier with the maximum perfect rotation estimation prediction of accuracy. The SVM training and classification algorithm for spam emails is offered within the set of rules under:

Algorithm 5.
Step 1. Input Sample Email Message x to order
Step 2. A preparing set S, a piece work, $\{c1, c2, \dots c_n\}$ and $\{\gamma1, \gamma2, \dots \gamma_n\}$.
Step 3. Number of closest neighbours k.
Step 4. for I = 1 to n
Step 5. set c=ci;
Step 6. for j = 1 to q
Step 7. set $\gamma=\gamma$;
Step 8. Produce a prepared SVM classifier f (x) fuzzy the current merger
       boundary (C, γ);
Step 9. if (f (x) is the first created discriminant work) at that point
Step 10. keep f (x) as the best SVM classifier f*(x);
Step 11. else
Step 12. Compare classifier f (x) and the current best SVM classifier f*(x)
       utilizing k-crease cross-approval
Step 13. keep class1.
Step 14. end if

Step 15. end for
Step 16. end for
Step 17. Return Final Email Message Classification (Spam/Non-spam email)
Step 18. end

15.4.7 Decision Tree

A Decision Tree (DT) is such a classifier whose model seems like that of a tree structure. Choice Tree presentation is an undeniable technique that prompts getting data on gathering. Each hub of a DT is either a leaf hub that decides the assessment of the normal segment (class). It can similarly be a choice hub that demonstrates certain test to be coordinated on the assessment of a component, with one branch and a sub-tree (which is a subset of the greater tree) addressing every conceivable eventual outcome of the test. A choice tree can be used to offer response for course of action issue by beginning at the establishment of the tree and encountering it until it gets to a leaf hub that gives the arrangement result. Choice Tree learning is a philosophy that has been applied to spam separating. The fact of the matter is to make a DT model and train the model with the objective for it to guess the assessment of a target variable zeroed in on different data factors. The specific inner hub talks with a part of the information variable. Singular leaf implies an assessment of the target variable gave that the assessments of the data factors are from the way that leads from the root to the leaf. It is possible to get comfortable with a tree by breaking the chief set into different subsets depending upon the assessment of the component that was given beforehand. This technique is iterated for each resultant subset more than once which propose the clarification it is known as recursive dividing. The recursion stops once all the subsets at a particular hub all have target factors that are tantamount. Another standard that can incite the finish of the recursion is while isolating the set is no all the additionally updating the gauges. There are different sorts of decision tree as explained below.

15.4.7.1 NBTree Classifier

 NBTree is a type of decision tree that hybridized Decision Tree with Naïve Bayes classifier where the strengths of each the algorithms are combined. This methodology works through making use of the NBTree Classifier at nodes whilst the selection tree is created with one variable that is separated at every node. For a database that is huge in size, the NBtree classifier is beneficial, if the size of the database is non-uniform and the highlights aren't inevitably self-ruling, the first-rate of the NBtree receives conspicuous.

The database of the unsolicited mail emails follows the above-portrayed example.

15.4.7.2   C4.5/J48 Decision Tree Algorithm

J48 is an adjusted, reordered and openly accessible form of C4.5 choice tree calculation. J48 is created by considering information at the hubs which are utilized to analyse the importance of common characteristics. By the choice tree a tree model is created the fuzzy the utilization of each component in turn. To rework the dataset, the calculation utilizes the estimation of the element. What's more, continue to look for the regions of the dataset that clearly have one class and show those regions as leaves. For the rest of the territories that contain classes that are multiple, the calculation chooses elective highlights. It likewise keeps up the separating cycle with simply the quantity of events in such regions forthcoming the time that the leaves are totally made, or there is nonappearance of highlight that can be used to make at any rate one leave fluctuated in the differ regions. The choice tree created by C4.5 can be applied for tackling distinctive order issues. The calculation chooses the highlights that it can additionally isolate into subclasses at every hub. The yield of the characterization acquired is signified by a leaf hub.

15.4.7.3   Logistic Version Tree Induction (LVT)

LVT is a Decision tree that uses logistic regression fashions on leaf nodes. LVT classifier has proven better degree of precision and sturdiness in numerous instructional regions. The logistic version is simple to decode in comparison to c4. 5 timber. Besides, it's been confirmed that LVT timber is compressed in size compared to the trees created by c4. Five initiations. The famous iterative dichotomized three (id3) set of rules proposed with the aid of ross Quinlan became discussed to construct the selection tree the use of entropy and facts benefit. The entropy evaluates the adulteration of a random series of email samples at the same time as the information benefit is used to calculate entropy by way of dividing the e-mail pattern with the aid of some functions. Assuming we've got an electronic mail dataset e with classifications CJ, entropy has calculated the use of eq. (15.15) below.

$$\text{entropy}\,(E) \sum_{j=1}^{|c|} \Pr(c_j)\log_2 \Pr(C_j) \qquad (15.15)$$

The connection between the information gain and entropy is represented in eq. (15.16) beneath.

$$\text{gain } (E, F_i) = \text{entropy } (D) - \text{entropy}_{Fi} (E) \qquad (15.16)$$

Algorithm 6. Decision tree set of rules for unsolicited mail filtering
Step 1. Enter email message dataset
Step 2. Calculate entropy for dataset
Step 3. At the same time as situation do
Step 4. For each attribute
Step 5. Compute entropy for all categorical values
Step 6. Take average records entropy for present characteristic.
Step 7. Calculate advantage for the current characteristic
Step 8. Select the best benefit characteristic
Step 9. Give up for
Step 10. Cease even as
Step 11. Go back very last e-mail message class (junk mail/non-junk mail e-mail)
Step 12. Quit

15.4.8   Ensemble Classifiers

Ensemble Classifiers is a brand-new approach in which a set of different classifiers are skilled and assembled to in addition enhance the class accuracy of the complete gadget on the same hassle, in this example it's for unsolicited mail filtering. They're a category of gadget studying algorithm that paintings in settlement and are carried out to enhance the class overall performance of the entire gadget. The most broadly universal ensemble classifiers are bagging and boosting. Those algorithms teach classifier instances on diverse subsets of the whole information set. Bagging combines the outputs of skilled classifiers on sample drawn from a larger pattern of the data set.

Boosting is a very green approach that combines a chain of "weak" newcomers to create an unmarried learner this is more potent than the man or woman learner. It is categorized as a getting to know the set of rules that are centered on the idea of hybridisation of numerous vulnerable hypotheses, a excellent instance is an AdaBoost device. The aim of boosting is to obtain a totally correct class rule through combining several susceptible rules or susceptible hypotheses every of which can be most effective exceedingly accurate.

Today, boosting is now applied in the field of category, regression, face recognition and so forth. Boosting algorithms that utilised confidence rated projections are being implemented to remedy unsolicited mail filtering problem. Due to its remarkable overall performance in fixing classification issues, Adaboost is widely normal gadget getting to know set of rules. Adaboost is fast, the set of rules is straightforward and clean to program, absence of parameter tuning (except t) makes is much less bulky.

15.4.9   Random Forests (RF)

The random forest is an example of a collaborative studying technique and regression method suitable for solving troubles that pertain to classifying records into companies. The fuzzy the use of decision trees this algorithm includes out estimation. These decision trees are then utilized for the venture of predicting the institution; this is executed by using deliberating the selected groups of each distinct timber and the institution that have the very best range of vote is taken as the end result. This method has gained recognition in recent times and it has located software in unique fields and in literature it has been used to offer an approach to similar trouble.

The strengths of the RF set of rules are that it typically has less sorting mistakes and greater f-scores compared to choice trees. Despite the fact that they are significantly less difficult to realize for humans its overall performance is commonly as much as or even higher than that of SVMs. Its performance is truly excellent with uneven statistics units that are taken into consideration via some missing variables. It presents an efficient mechanism for computing the approximate value of missing data and maintaining precision in conditions in which a huge percentage of the records are lost. RF allows the consumer to grow as many trees as possible.

The rate of its execution is high. In numerous instances wherein information set size is big, a good deal of memory for the storage of the facts is needed. Computing the closeness shows that an increase inside the storage space needed is without delay proportional to the number of times accelerated with the aid of the number of bushes. The project of classifying new records from an enter vector starts by way of putting the input vector alongside each of the bushes in the forest. Each tree will carry out its category that's frequently mentioned as the tree "votes" for that institution. The woodland chooses which of the agencies have the overall maximum votes inside the wooded area.

15.5   Conclusion

In this part of the discussion, we studied gadget gaining knowledge of methods and their software in the context of junk mail filtering. An assessment of numerous algorithms has been applied for the class of messages as either junk mail or ham as furnished. The attempts made by researchers in solving the problem of spam emails fuzzy gadget gaining knowledge of classifiers are mentioned. The evolution of junk mail messages over time to steer clear of filters is likewise observed.

The obvious structural layout of an email junk mail filter and the tactics used in filtering unsolicited emails has been noted. This chapter plotted a number of the public datasets and metrics that can be used to measure the effectiveness of any unsolicited mail clear out.

The challenges of the machine learning algorithms in proficiently handling the chance of junk mail has been pointed out and relative studies of the machine learning technics available in the literature have been completed. Additionally, a few open studies issues associated with junk mail filters have been explored. The literature we reviewed shows that substantial progress has been made and still is being made in the subject. Because of the open issues in junk mail filtering, further research to evaluate the effectiveness of junk mail filters need to be done. This could make the increase of junk mail filters an important investigation subject for academicians and industry practitioners.

References

1. Cárdenas, A. A., Berthier, R., Bobba, R.B., Huh, J.H., Jetcheva, J.G., Grochocki, D., & Sanders, W.H. (2014) "A Framework for Evaluating Intrusion Detection Architectures in Advanced Metering Infrastructures," *IEEE Transactions on Smart Grid*, vol. 5(2), pp. 906–915.
2. Friedman, R. W., & Schuster. A. (2008) "Providing kAnonymity in Data Mining," *VLDB Journal*, vol. 17(4), pp. 789–804.
3. Singh, R., Kumar, P. & Diaz, V. (2020) "A Holistic Methodology for Improved RFID Network Lifetime by Advanced Cluster Head Selection using Dragonfly Algorithm," *International Journal of Interactive Multimedia and Artificial Intelligence,* vol. 6(2), pp. 8.
4. Singh, B., Singh, R. & Rathore. P.S. (2013) "Randomized Virtual Scanning Technique for Road Network," *International Journal of Computer Applications*, vol. 77(16), pp. 1–4.
5. Kumar, N., Triwedi, P. & Rathore, P.S. (2018) "An Adaptive Approach for image adaptive watermarking using Elliptical curve cryptography (ECC)"

First *International Conference on Information Technology and Knowledge Management*, pp. 89–92, ISSN 2300-5963.

6. Bhargava, N., Singh, P., Kumar, A., Sharma, T. & Meena, P. (2017) "An Adaptive Approach for Eigenfaces-based Facial Recognition" *International Journal on Future Revolution in Computer Science & Communication Engineering (IJFRSCE)*, vol. 3(12), pp. 213–216.

7. Herzberg, A. & Gbara, A. (2004) "Trustbar: Protecting (even naive) Web Users from Spoofing and Phishing Attacks," *Cryptology* ePrint Archive Report, pp. 155.

8. Rathore, P., S., Chaudhary A. & Singh, B. (2013) "Route planning via facilities in time dependent network," *IEEE Conference on Information & Communication Technologies*, pp. 652–655.

9. Fu, A. Y,, Wenyin, L. & Deng X (2006) "Detecting Phishing Web Pages with Visual Similarity Assessment Based on Earth Mover's Distance (emd)," *IEEE Transactions on Dependable and Secure Computing*, vol. 3(4), pp. 301–311.

10. Manek, A., S., Shenoy, P., D., Mohan, M., C. & Venugopal K R, (2016) "Detection of Fraudulent and Malicious Websites by Analysing User Reviews for Online Shopping Websites," *International Journal of Knowledge and Web Intelligence*, vol. 5(3), pp. 171–189.

11. Wu, B., Lu, T., Zheng, K., Zhang, D. & Lin, X. (2015) "Smartphone Malware Detection Model Based on Artificial Immune System," *China Communications*, vol. 11(13), pp. 86–92.

12. Dwork, C., McSherry, F., Nissim, K. & Smith, A. (2006) "Calibrating Noise to Sensitivity in Private Data Analysis," *Theory of Cryptography Conference*, pp. 265–284.

13. Jackson, C., Simon, D.R., Tan, D. S. & Barth, A. (2007) "An Evaluation of Extended Validation and Picturein-Picture Phishing attacks," *International Conference on Financial Cryptography and Data Security*, pp. 281–293.

14. Rathore, P.S. (2017) "An adaptive method for Edge Preserving Denoising, International Conference on Communication and Electronics Systems, Institute of Electrical and Electronics Engineers, *Proceedings of the 2nd International Conference on Communication and Electronics Systems (ICCES 2017)*.

15. Tseng, C., Y., Balasubramanyam, P., Limprasittiporn, R., Rowe, J. & Levitt, K. (2016) "A Specification-Based Intrusion Detection System" *Global Journal of Computer Science and Technology*, vol. 16(5), pp. 125–134.

16. Beaver, D., Micali, S. & Rogaway, P. (1990) "The Round Complexity of Secure Protocols," *Proceedings of the 22nd Annual ACM Symposium on Theory of Computing*, pp. 503–513.

17. Bhargava, N., Dayma, S., Kumar, A. & Singh, P. (2017) "An approach for classification using simple CART algorithm in WEKA," *11th International Conference on Intelligent Systems and Control (ISCO)*, pp. 212–216.

About the Editors

Neeraj Bhargava, PhD, is a professor and head of the Department of Computer Science at Maharshi Dayanand Saraswati University in Ajmer, India, having earned his doctorate from the University of Rajasthan, Jaipur in India. He has over 30 years of teaching experience at the university level and has contributed to numerous books throughout his career. He has also published over 100 papers in scientific and technical journals and has been an organizing chair on over 15 scientific conferences. His work on face recognition and fingerprint recognition is often cited in other research and is well-known all over the world.

Ritu Bhargava, PhD, is an assistant professor in the Department of Computer Science at Sophia Girls College in Ajmer, India, having earned her PhD in computer science from Hemchandracharya North Gujarat University Patan, Gujarat, India. She has more than 15 years of active teaching and research experience and has contributed to three books and more than 30 papers in scientific and technical journals. She has also been an organizing chair on over 15 scientific conferences, and, like her colleague, her work on face recognition and fingerprint recognition is well-known and often cited.

Pramod Singh Rathore, MTech, is an assistant professor at the Aryabhatta College of Engineering and Research Center and visiting faculty member at MDSU in Ajmer, India. He is a PhD in computer science and engineering at the University of Engineering and Management and already has eight years of teaching experience and over 45 papers in scientific and technical journals. He has also co-authored and edited numerous books.

Rashmi Agrawal, PhD, is a professor in the Department of Computer Applications at the Manav Rachna International Institude of Research and Studies in Faridabad, India with more than 18 years of teaching experience. She is a book series editor and the associate editor on a scientific journal on data science and the internet of things. She has published many research papers in scientific and technical journals in these areas and contributed multiple chapters to numerous books. She is currently guiding PhD students and is an active reviewer and editorial board member of various journals.

Index